Kinematics of Machinery Through HyperWorks

HISTORY OF MECHANISM AND MACHINE SCIENCE

Volume 18

Series Editor

MARCO CECCARELLI

Aims and Scope of the Series

This book series aims to establish a well defined forum for Monographs and Proceedings on the History of Mechanism and Machine Science (MMS). The series publishes works that give an overview of the historical developments, from the earliest times up to and including the recent past, of MMS in all its technical aspects.

This technical approach is an essential characteristic of the series. By discussing technical details and formulations and even reformulating those in terms of modern formalisms the possibility is created not only to track the historical technical developments but also to use past experiences in technical teaching and research today. In order to do so, the emphasis must be on technical aspects rather than a purely historical focus, although the latter has its place too.

Furthermore, the series will consider the republication of out-of-print older works with English translation and comments.

The book series is intended to collect technical views on historical developments of the broad field of MMS in a unique frame that can be seen in its totality as an Encyclopaedia of the History of MMS but with the additional purpose of archiving and teaching the History of MMS. Therefore the book series is intended not only for researchers of the History of Engineering but also for professionals and students who are interested in obtaining a clear perspective of the past for their future technical works. The books will be written in general by engineers but not only for engineers.

Prospective authors and editors can contact the series editor, Professor M. Ceccarelli, about future publications within the series at:

LARM: Laboratory of Robotics and Mechatronics
DiMSAT – University of Cassino
Via Di Biasio 43, 03043 Cassino (Fr)
Italy
E-mail: ceccarelli@unicas.it

For other titles published in this series, go to
www.springer.com/series/7481

J. S. Rao

Kinematics of Machinery Through HyperWorks

Springer

J.S. Rao
Altair Engineering
Chief Science Officer
Outer Ring Road
560103 Bangalore
India
js.rao@altair.com

Additional material to this book can be downloaded from http://extras.springer.com

ISSN 1875-3442 e-ISSN 1875-3426
ISBN 978-94-007-1155-6 e-ISBN 978-94-007-1156-3
DOI 10.1007/978-94-007-1156-3
Springer Dordrecht Heidelberg London New York

Library of Congress Control Number: 2011922909

Cover design: SPi Publisher Services

Printed on acid-free paper

Springer is part of Springer Science+Business Media (www.springer.com)

Dedicated to the memory of my parents

Jammi Chikka Rao
Jammi Ramanamma

Contents

Preface

The Theory of Machines was borne as a subject with the Industrial Revolution and the birth of Reciprocating Steam Engine nearly 230 years ago. The reciprocating steam engine was the main work horse for just over hundred years during the 19th century and gradually lost its place in history by the turn of 20th century. It is then the turn of Internal Combustion Engines and Rotating Machinery.

Kinematics and Dynamics of Reciprocating and Rotating machinery is the fundamental study to mechanical engineers before proceeding to stress design. These analyses were done mainly by graphical methods for planar mechanisms as they gave good insight into the mechanism and repetitive history of velocities, accelerations, static and dynamic forces, etc.

All machines are periodic in operation depending on thermodynamic cycles for example one revolution in a two-stroke engine or two revolutions in a four-stroke engine. Besides thermodynamics as a fundamental mechanical engineering subject, Theory of Machines is the first subject a fresh entrant to mechanical engineering studies faces. The concepts of moving machine members over a period of thermodynamic cycle and the variation of displacements, velocities and accelerations form the subject of Kinematics. Here we do not question what makes these machine members move but merely find the kinematics when there is a motion. When we ask the question of forces that make the motion, we are dealing kinetics; together we have Dynamics of Machinery. When we include the machinery aspects such as links, kinematic chains, mechanisms, etc., to form a given machine we have the subject of Theory of Machines.

Usually this subject is introduced as a two-semester course, kinematics and kinetics simultaneously with Thermodynamics or Heat Engines before the design of machine members begins. This book forms the material for first semester of Theory of Machines.

As this subject is over 200 years old, there are several textbooks already available. The new books that appear from time to time take into account new techniques available; the subject matter itself has not changed particularly for the entry level. What is the difference here then?

This book attempts to bring in the machine live on to the screen and explain the theory of machines concepts through animations and introduce how the problems are solved in industry to get complete history in the shortest possible time rather than using graphical (or analytical) methods that are in vogue even today. Thus the student is introduced to the concepts through visual means and brings him close to industrial applications by the end of the two semester program taking him well equipped for design courses.

International Federation for the Promotion of Mechanism and Machine Science (IFToMM) has developed a standard nomenclature and notation on Mechanism and Machine Science and this book adopts these standards so that any communication between scientists and teachers in classrooms across the world can be with the same terminology causing no confusion.

This book adopts HyperWorks MotionSolve to perform the analysis and visualizations, though the book is independent of the requirement of any software. Having this software helps in further studies and analysis. The avis in this book can be accessed from extras.springer.com by using the ISBN; they are: Figs. 2.1, 2.3, 2.4, 2.5, 2.6, 2.7, 2.8, 2.9, 2.10, 2.11b, 2.12a, 2.12b, 2.13, 2.20, 2.21, 2.22, 2.23, 2.24, 2.25, 2.26, 2.27, 2.28, 2.29, 2.30, 2.31, 2.32, 2.33, 3.1, 3.2a, 3.2b, 3.2c, 3.5a, 3.13f, 3.20, 3.28, 3.30, 3.38, 4.1, 4.2, 4.3, 4.4, 4.5,4.6, 4.8, 4.9, 4.10, 4.11, 4.12a, 4.12b, 4.12c, 4.13, 4.14, 4.15,4.16, 4.18, 4.19a, 4.19b, 4.20, 4.22a, 4.22b, 4.22c, 5.1, 5.2, 5.3, 5.4, 5.5a, 5.5b, 5.6a, 5.6b, 5.6c, 5.7, 5.17, 5.18a, 5.18b, 5.26, 5.29, 5.30, 5.31, 5.32-33, 5.34-35, 6.1, 6.2a, 6.5a, 6.7b, 6.10, 6.18a, 6.18b, 6.23a, 6.23b, 6.23c, 6.24a, 6.24b, 7.4c, 7.4d and 8.1.

The author acknowledges help given by various students and colleagues over four decades. Of particular mention, I thank Professor J. Srinivas, Anil Sakhamuri, Uday M. Udapi and Sundar Nadimpalli. I am also thankful to Mr. Pavankumar and Mr. Nelson Dias of Altair Engineering India.

Finally, I am ever so thankful to my beloved wife Indira for her understanding in my work and cooperation.

Chapter 1
Beginnings of the Theory of Machines[1]

It is always fascinating to know the origins of any subject. The subject of Theory of Machines began during the era of James Watt with the Industrial Revolution. We will look at what happened prior to this.

1.1 Beginning of the Wheel

During the Mesolithic, or Middle Stone Age, some many thousands of years ago, man found that a section of a tree trunk could be moved more easily under the force of gravity because it was round. If the branches and twigs of the trunk were removed, the speed of the rolling log improved.

Early men began to place runners under a heavy load, which they discovered would make it easier for the load to drag. This was the invention of the sledge. Men then began to combine the roller and the sledge. As the sledge moved forward over the first roller, a second roller was placed under the front end to carry the load when it moved off the first roller. It was discovered that the rollers which carried the sledge became grooved with use and that these deep grooves actually allowed the sledge to advance a greater distance before the next roller was needed to come on. Thus, the rollers were changed into wheels. In the process of doing so, sections of wood between the grooves of the roller were cut away to form an axle and wooden pegs were fastened to the runners on each side of the axle. A slight improvement was made to the cart. This time, instead of using pegs to join the wheels to the axle, holes for the axle were drilled through the frame of the cart. Axle and wheels were now made separately.

The wheel is probably the most important mechanical invention of all time. Nearly every machine built since the beginning of the industrial revolution involves a single, basic principle embodied in one of mankind's truly significant inventions.

[1] This chapter is based on the paper "History of Rotating Machines", *IFToMM Workshop on the History of Machines and Mechanisms, HMM 2007*, Bangalore, December 14, 2007.

It is hard to imagine any mechanized system that would be possible without the wheel or the idea of a symmetrical component moving in a circular motion on an axis. From tiny watch gears to automobiles, jet engines and computer disk drives, the principle is the same.

Agricultural villages had begun to develop by 8000 BC. This is known as the Neolithic period, or New Stone Age. During this time the slow potter's wheel was invented. In about 3000 BC Egyptians developed the fast wheel, a completely mobile, carefully balanced apparatus of stone. Based on diagrams on ancient clay tablets, the earliest known use of this essential invention was a potter's wheel that was used at *Ur* in Mesopotamia (part of modern day Iraq) as early as 3500 BC. The first use of the wheel for transportation was probably on Mesopotamian chariots in 3200 BC. It is interesting to note that wheels may have had industrial or manufacturing applications before they were used on vehicles.

The wheel was furthered improved on later by the Egyptians, who made wheels with spokes, which could be found on Egyptian chariots of around 2000 BC. Over in Ancient India, chariots with spoked wheels dating back to around 1500 BC were also discovered. The Greeks too, adopted the idea of wheel-making from the Egyptians and made further improvements to it. Later, during the time of the Roman Empire, the Romans too engaged themselves in wheel-making and produced the greatest variety of wheeled vehicles. They had chariots for war, hunting, and racing, two-wheeled farm carts, covered carriages, heavy four-wheeled freight wagons and passenger coaches.

With the collapse of the Roman Empire in AD 476, the wheel became widely used for war machines across the old empire. The grinding wheel was introduced from Arabia to Europe in the middle ages, greatly improving the effect of bladed combat weapons.

1.2 Archimedes (287–212 BC)

The first to systematize the simple machines and propound the theory of their functions was Archimedes of Syracuse in Sicily. It was probably he who invented the compound pulley, a device for increasing traction or lifting power and he propounded the theory of the lever, both one- and two-armed. He regarded the wheel as a circular figure described by a rotating one-armed lever, and the screw as the circular analogy of the inclined plane. One of his famous sayings is "Give me a place to stand and I will move the earth".

Archimedes received his education at the University of Alexandria, where groups of mathematicians and scientists worked, devoting themselves to the construction of numerous fascinating machines. The greatest and most colorful of what is known as the Alexandrian school of engineers was undoubtedly Hero who lived sometime during the second century BC. His best invention was the *aelopile*, the first reaction turbine, which converted heat into mechanical energy through the medium of steam.

Hero's *aelopile*, the first reaction turbine, could not produce useful work, as the speed was not sufficient to create the required high head of steam. In the 1780s James Watt worked on the theoretical operating conditions of a reaction turbine and he concluded that such a turbine could not be built given the state of contemporary technology.

1.3 Water Wheels

In all likelihood, the earliest tools employed by humankind for crushing or grinding seeds, nuts, and other food-stuffs consisted of little more than a flat rock, upon which the material was crushed by pounding with a stone or tree branch. The archaeological records show that as early as 30,000 years ago, Cro-Magnon artists employed the mortar and pestle to grind and mix the pigments they used to create their magnificent "cave-art".

Far more efficient than the flat rock or even the mortar and pestle was the handmill, which appears to have long pre-dated the agricultural revolution. The handmill consists of a flat rock, often hollowed or concave, on which the grain, seeds, or other materials is placed, and a grinding stone, which is rolled across the grain, thus reducing the grain to flour. Although the handmill is still, today, in use in many parts of the world, approximately 2,000 years ago humankind began to harness waterpower to turn the stones that ground its grain. They were probably the first tools for creating mechanical energy that replaced humans and animals.

The first description of a water wheel is from Vitruvius, a Roman engineer (31 BC–14 AD), who composed a 10 volume treatise on all aspects of Roman engineering. From classical times, there have existed three general varieties of water wheels: the horizontal wheel and two variations of the vertical wheel.

Waterpower was an important source of energy in ancient Chinese civilization. One of the most intriguing applications was for iron casting. According to an ancient text, in 31 AD the engineer Tu Shih invented a water-powered reciprocator for the casting of (iron) agricultural implements. Waterpower was also applied at an early date to grinding grain.

Renaissance engineers studied the waterwheel and realized that the action of water on a wheel with blades would be much more effective if the entire wheel were somehow enclosed in a kind of chamber. They knew very well that only a small amount of the water pushing or falling on a wheel blade or paddle actually strikes it, and that much of the energy contained in the onrushing water is lost or never actually captured.

1.4 Wind Mills

Over 5,000 years ago, the ancient Egyptians used wind to sail ships on the Nile River. While the proliferation of the water mill was in full swing, wind mills appeared to harness more inanimate energy by employing wind sails. Prototypes of wind mills were probably known in Persia (present day Iran) as early as 7th century AD with the sails mounted on a vertical axis. Towards the end of the 12th century, wind mills with sails mounted on a horizontal axis appeared in Europe; the first of this kind probably appeared in Normandy, England. These are post mills, where the sails and machinery are mounted on a stout post and the entire apparatus has to be rotated to face the wind.

Two centuries later the tower mill was introduced, enclosing the machinery in a stationary tower so that only the cap carrying the sails needed to be turned to the wind.

In 1854 Daniel Halliday obtained the first American windmill patent. His windmill had four wooden blades that pivoted and would self adjust according to wind speed. It had a tail which caused it to turn into the wind.

1.5 Renaissance Engineers

The credit for making pressure exerted by the atmosphere entirely explicit belongs to Otto von Guericke, who in 1672 published the famous book in which he described his air pump and the experiments that he made with it from the mid 1650s onwards. Once it was understood that atmosphere exerts pressure, it was a matter of creating a vacuum and utilizing atmospheric pressure to move the piston in a cylinder.

Denis Papin (1647–1712) a French physicist, mathematician and inventor is best known for his pioneering invention of the steam digester, the forerunner of the steam engine. He visited London in 1675, and worked with Robert Boyle from 1676 to 1679, publishing an account of his work in *Continuation of New Experiments* (1680). During this period, Papin invented the *steam digester,* a type of pressure cooker. He first addressed the Royal Society in 1679 on the subject of his digester, and remained mostly in London until about 1687, when he left to take up an academic post in Germany. While in Leipzig in 1690, having observed the mechanical power of atmospheric pressure on his 'digester', he built a model of a piston steam engine, the first of its kind.

Thomas Savery (1650–1715) was an English military engineer and inventor who in 1698 patented the first crude steam engine, based on Denis Papin's Digester or pressure cooker of 1679. His machine consisted of a closed vessel filled with water into which steam under pressure was introduced. This forced the water upwards and out of the mine shaft. Then a cold water sprinkler was used to condense the steam. This created a vacuum which sucked more water out of the mine shaft through a bottom valve.

In 1705 Papin developed a second steam engine, with the help of Gottfried Leibniz, using steam pressure rather than atmospheric pressure. Papin's steam engine was a breakthrough since Hero's reaction turbine of the second century BC never functioned in reality.

The Newcomen steam engine was the first practical device to harness the power of steam to produce mechanical work. Newcomen's first working engine was installed at a coal mine at Dudley Castle in Staffordshire in 1712. They were used throughout England and Europe to pump water out of mines starting in the early 18th century and were the basis for James Watt's later improved versions. Although Watt is far more famous today (due largely to Matthew Boulton's tireless salesmanship), Newcomen rightly deserves the majority of the credit for widespread introduction of steam power.

1.6 Industrial Revolution

Between 1780 and 1850, in a space of just seven decades, the face of England was changed by a far-reaching revolution, without precedent in the history of mankind.

Glasgow University had one of the Newcomen engines for its natural philosophy class. In 1763, one hundred years after the birth of Newcomen, this apparatus went out of order and Professor John Anderson gave the opportunity to James Watt (1736–1819) to repair it. After the repair and while experimenting with it, he was struck by the enormous consumption of steam; at every stroke, the cylinder and piston had to be heated to the temperature of boiling water and cooled again. This prevented the apparatus from making, with the available boiler capacity, more than a few strokes every minute. He quickly realized that the wastage of steam is inherent in the design of the engine and became obsessed with the idea of finding some remedy. From the discovery of Dr. Joseph Black (1728–1799), he deduced that the loss of latent heat was the most serious defect in the Newcomen engine. The work of James Watt is thus the application of science to engineering that led to the birth of industrial revolution.

In 1765 he conceived the idea of a separate condensing chamber for the steam engine to separate the condensation system from the cylinder, injecting the cooling water spray in a second cylinder, connected to the main one. When the piston had reached the top of the cylinder, the inlet valve was closed and the valve controlling the passage to the condenser was opened. External atmospheric pressure would then push the piston towards the condenser. Thus the condenser could be kept cold and under less than atmospheric pressure, while the cylinder remained hot. Important as the separate condenser idea was, in the fully developed version of 1775 that went into production, changes had to be more far-reaching. There was no spray, the condenser being immersed in a water tank and at each stroke the warm condensate was drawn off and sent up to a hot well by a vacuum pump which also helped to evacuate the steam from under the power cylinder. The still-warm condensate was recycled as feed water for the boiler.

Reciprocating machinery has inherent disadvantages at high speeds; they have practically disappeared in the modern day world. There are still steam locomotives operating in a few places, e.g., Fairy Queen, the oldest running vintage steam locomotive in the world built in the year 1855 by the British firm Kinston, Thompson & Hewitson for the British firm East India Railways and occasional reciprocating engines for producing small amounts of power in sugar mills, but otherwise they are gone. Internal combustion engines still thrive for transportation, power generation, and so on.

1.7 The Nature of This Book

The subject Theory of Machines is about 200 years old and has undergone tremendous changes during these two centuries. It began with the need to develop understanding of various links of disparate mechanisms, followed by Kinematics that explains displacements, velocities and accelerations inherent in these mechanisms. The reciprocating steam engines were bulky and rigid, their speeds were very low and rarely was a dynamic analysis needed. Their loads were calculated from dead weights and statics and applied to the pursuit of better design. By the time the basic concepts of Dynamics were perceived, the reciprocating steam engine was already on its way out, with about a century left to go, and Reciprocating Internal Combustion engines gained importance. Simultaneously, rotating steam turbines were favored for high capacities and high speeds. Jet engines and Gas turbines came into vogue nearly six decades ago. These new devices significantly enriched the subject Theory of Machines. Dynamics assumed an increasingly central role in design.

While reciprocating machinery formed the central theme of Kinematics and Dynamics studies, there were several other elements that became important in the study of the Theory of Machines, e.g., Cams in reciprocating engines, Gears in transmission units, Governors and Controls, etc. The studies of Reciprocating steam engine soon disappeared from the curriculum.

Initially Kinematics and Kinetics of Machinery was conducted by using graphical methods and where possible analytical methods until the advent of computer era in 1960s. With further advances in hardware and work centers the analysis and design was made simpler through simulation and visualization. Initially the designs were all through expensive and time consuming tests, they got gradually transformed to current day practice of simulation and optimization to arrive at the final prototype in the fastest time thus reducing the design cycle time and bringing new products to the market in the least possible time.

It is common practice nowadays for designers to use commercially available simulation tools or codes to achieve their finished designs. Most engineering instruction, however, continues to follow a conventional curriculum of classical theory, followed by computational tools, ending with amalgamation of this knowledge in order to meet industry standards. In this book, we attempt to make the time of instruction more compact and to bring engineering students into the design industry in a faster and more efficient manner.

For an engineer to be most effective for current day applications, it is necessary to be well versed in the subject and have the ability to apply the basics to achieve advanced designs using tools that are fast acting and efficient. Therefore the Theory of Machines, taught usually in two semesters, is proposed to be given through simulation tools to make a young engineer most efficient in a shorter period of time. The two semester pattern can continue with additional tasks of practicing applications of industrial nature. To assist in this model of instruction and learning, the subject matter is developed here through Altair MotionSolve and MotionView. The teacher can instruct students in the basics of these tools to allow the students have imagination and develop industrial practice skills while learning the basics of the subject.

Our approach to such instruction is to teach the visualization of moving members of a machine, and correlation of these motions, to determine the most important kinematic parameters in their design. We see it as a more effective path than imagining a motion period through several 2D figures of the machine. We believe that it will more effectively help the beginning engineer understand the function and analysis of a reciprocating engine, cams that open and close valves, gears that produce uniform angular velocities between two shafts in a transmission unit, governors, links that provide reclining and movement of seats, earth moving elements handling large quantities of earth material, special purpose machine tools as in bottling systems, robots, etc., all machinery members and functions that one can imagine would have applications of Theory of Machines.

We will first look at kinematics and later extend our discussion to kinetics. The subject of kinetics will focus on important aspects of forces acting on machine members that will subsequently need to be recognized in design.

Note that this is not an instruction manual for any commercial tool. An appropriate manual can be given out separately and the teacher can, if it seems desirable, simultaneously teach the steps that will enable students to most effectively follow the subject matter through visualization. While the time-honored and conventional practice of drawing outlines and specifications on drafting tables will undoubtedly be quickly discouraged, students should be encouraged to do free hand sketching to express their ideas, particularly at the concept stage.

Chapter 2
Planar Mechanisms

Every machine has moving members to perform a designed job. There are Prime Movers or Drives that use energy from some source to provide the necessary power to a Driven Machine or Load. The Prime Movers can be reciprocating steam engines (almost extinct now) which use steam from external combustion of coal in a boiler, reciprocating diesel or petrol engines which contain internal combustion in a cylinder, steam turbines using steam from external combustion of coal, gas turbines that use oil or gas in external combustion to produce hot gases or electrical machines that use electric power produced in power plants and transmitted to the location where it is needed. These prime movers drive a load that could be a propeller of a ship, an automobile for road transportation, a reciprocating or rotating compressor in oil and gas industries, an aircraft for air transportation, a generator to produce electricity or a machine tool that removes metal in a manufacturing process, etc.

As an example, in Figure 2.1a we show a Shaping Machine. The power from an electric motor is transferred through the crank, sliding block, rocker arm and connecting link to the cutting tool. These are rigid bodies to which the cutting tool is connected in a suitable manner to give constrained motion and remove the metal by shaping action. Theory of Machines is concerned with the layout of such members in a given machine, their motion and understanding of how the kinematic quantities, displacement, velocity and acceleration vary with time and kinetics of these members, viz., what forces and moments cause these motions. From these forces and moments, we can determine the stresses and size the members of the machine accordingly. That will be subject of Machine Design, which is not a part of Theory of Machines.

Before we proceed to formally develop the subject in a systematic manner, let us look at Figure 2.1b that depicts schematically the shaping machine of Figure 2.1a. The crank, sliding block, rocker arm, connecting link and the ram here are represented simply by lines with joints connecting them and provide a transfer of motion from the crank to the ram. Comparing Figures 2.1a and 2.1b we find the same motions are achieved by different bodies, e.g., watch the ram and the tool attached to

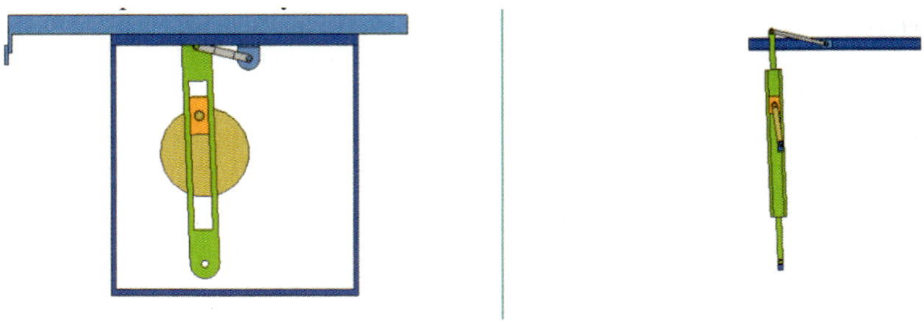

(a) (b)

Fig. 2.1 Shaping machine schematically represented

it.[1] Therefore we recognize that we can schematically represent each machine member by a rigid body. If all the bodies of the machine move in one plane as in this case, the rigid bodies are simply represented by straight line between two joints where the neighboring bodies of the machine are attached. The student should quickly learn to imagine what would be the actual shape of a machine component which might be schematically shown as a straight line (or as a triangle if there are three attachments to this body). Now let us define these machine components in a systematic manner to develop the subject further.

2.1 Basic Kinematic Concepts

The International Federation for the Promotion of Mechanism and Machine Science (IFToMM) is a world body established in 1969 for promotion of Mechanisms and Machines Science. One of its tasks is standardization of Mechanism and Machine Science terminology, the first step of which resulted in a concise publication in 2003 covering various aspects of Mechanisms and Machines. In this book, where necessary, all definitions are taken from these terminology standards.[2]

Machine

Mechanical system that performs a specific task, such as the forming of material, and the transference and transformation of motion and force. It comprises of several rigid bodies connected in such a way that it produces constrained relative motion between them and transmit forces and couples from the source of input power to result in motion, see Figures 2.1a and 2.1b.

[1] The avis in this book can be accessed from extras.springer.com.

[2] IFToMM Commission A, Terminology for the Mechanism and Machine Science, *Mechanism and Machine Theory*, Vol. 38, Nos. 7–10, pp. 598–1111, 2003.

Mechanism

System of bodies designed to convert motions of, and forces on, one or several bodies into constrained motion of, and forces on, other bodies. When we deal with a mechanism, the specific task to which it is assigned is not important, be it a shaping machine or an internal combustion engine, etc. The emphasis is on motion and force of the system and they may be applicable to any machine in which a specific mechanism may be employed. It is also a kinematic chain with one of its components (link or joint) connected to the fixed frame.

Planar Mechanism

Mechanism in which all points of its links describe paths located in parallel planes.

Link

Mechanism element (component) carrying kinematic pairing elements. Thus, the crank, sliding block, rocker arm, connecting link and the ram are the links of the shaping machine.

Bar

Link that carries only revolute joints.

Element (Pairing Element)

Assembly of surfaces, lines or points of a link through which it may contact some other link so constituting a kinematic pair. The connecting rod of an internal combustion engine is shown in Figure 2.2a. The big-end (one element) is connected to the crank-pin and the small end (another element) to the piston-pin. Since it is a rigid body, this link is usually represented as a line (as shown in Figure 2.2b) with its two end elements represented by small circles (joint). In other words, the part of a link that is connected to a neighboring link is called an element.

Joint

The physical realization of a kinematic pair.

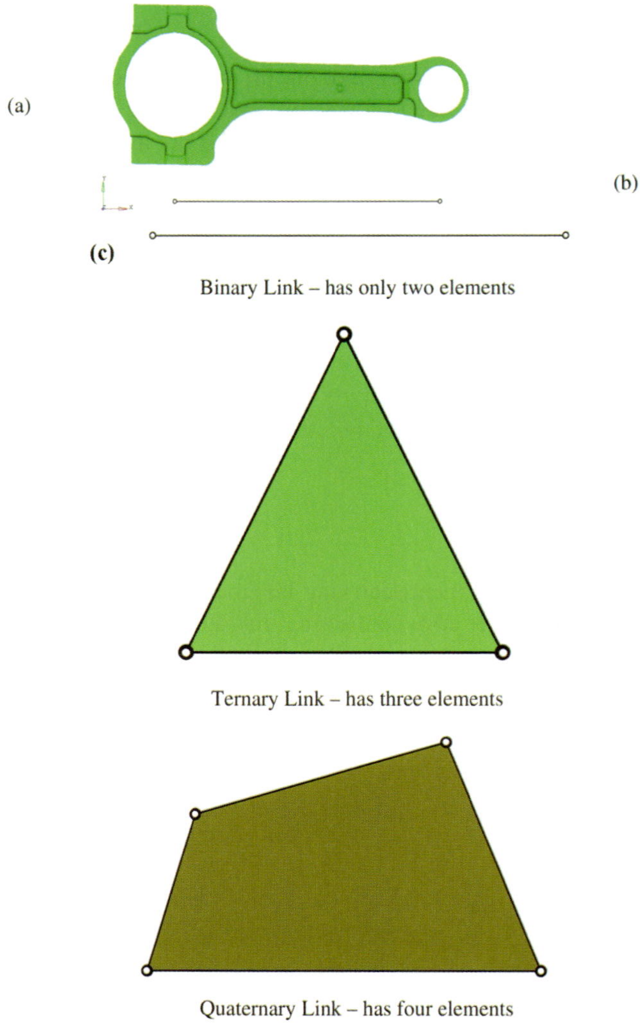

Fig. 2.2 (a) Con rod of an internal combustion engine actual rod represented by (b) a link; (c) different links

Kinematic Pair

Idealization of a physical joint that is concerned only with the type of constraint that the joint offers. There are many types of links, a few of which are shown in Figure 2.2c.

Kinematic Chain

Assemblage of links and joints.

Linkage

Kinematic chain whose joints are equivalent to lower pairs only.

Degree of Freedom [Connectivity]

The number of independent coordinates needed to describe the relative positions of pairing elements. A rigid body has six degrees of freedom as shown in Figure 2.3. Depending on the constraints imposed on the motion, the body may loose one or more of the six degrees of freedom.

Constraint

Any condition that reduces the degree of freedom of a system.

Closure of a Kinematic Pair

Process of constraining two rigid bodies to form a kinematic pair by force (force closure), geometric shape (form closure or self-closed), or flexible materials (material closure).

Force-Closed [Open] Pair

Kinematic pair, the elements of which are held in contact by means of external forces.

Form-Closed Pair [Self-Closed Pair]

Kinematic pair, the elements of which are constrained to contact each other by means of particular geometric shapes.

Figure 2.4 shows a sphere between two parallel plates giving five degrees of freedom for the kinematic pair. A circular cylinder between two parallel plates is shown in Figure 2.5, giving rise to a four degree of freedom kinematic pair.

Fig. 2.3 Rigid body with six degrees of freedom

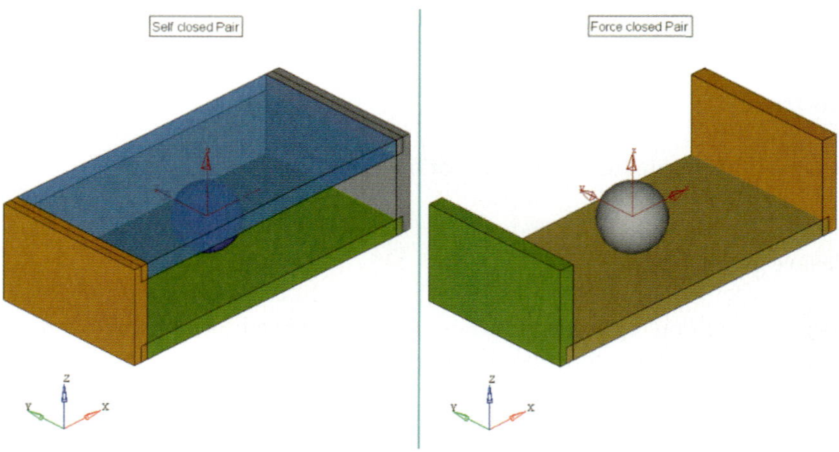

Fig. 2.4 A sphere between two parallel plates giving five degrees of freedom

Planar Contact [Sandwich Pair]

Pair for which the degree of freedom is three and that allows relative motion in parallel planes. A rectangular block between the two parallel plates in Figure 2.6 has only three degrees of freedom.

Spherical Pair

Pair for which the degree of freedom is three and that allows independent relative rotations about three separate concurrent axes.

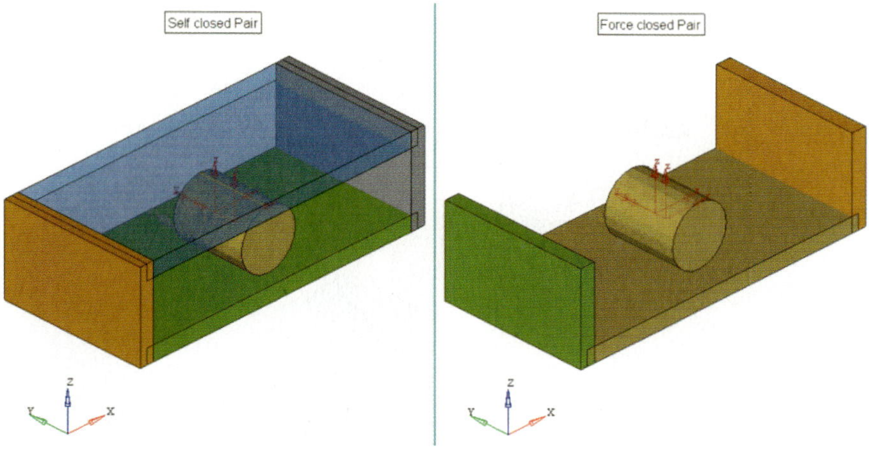

Fig. 2.5 A circular cylinder between two parallel plates with four degrees of freedom

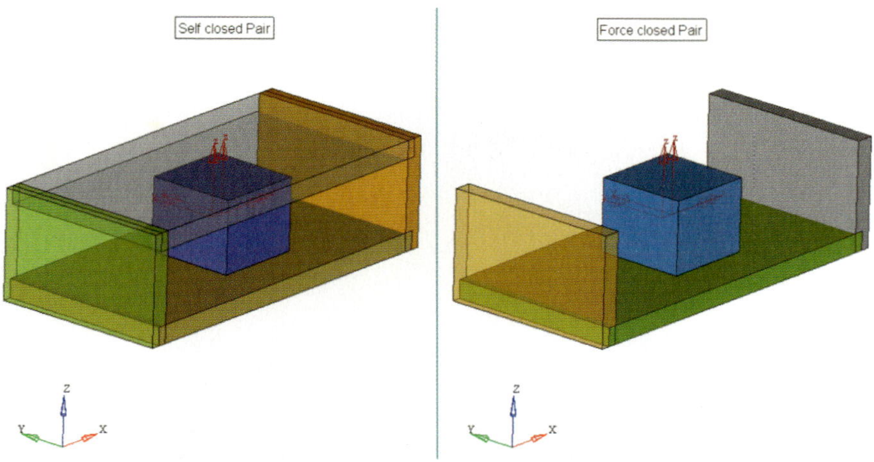

Fig. 2.6 A rectangular block between the two parallel plates has only three degrees of freedom

Cylindrical Pair

Pair for which the degree of freedom is two and that allows a rotation about a particular axis together with an independent translation in the direction of this axis. A round shaft in a coaxial cylinder in Figure 2.7 has only two degrees of freedom.

Turning Pair [Revolute Pair, Hinge]

Pair that allows only a rotary motion between its elements, see Figure 2.8.

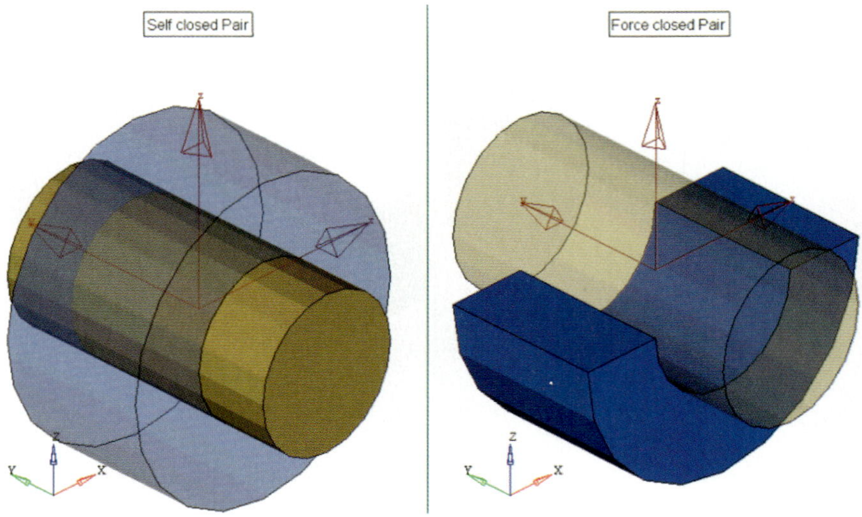

Fig. 2.7 A round shaft in a coaxial cylinder has only two degrees of freedom

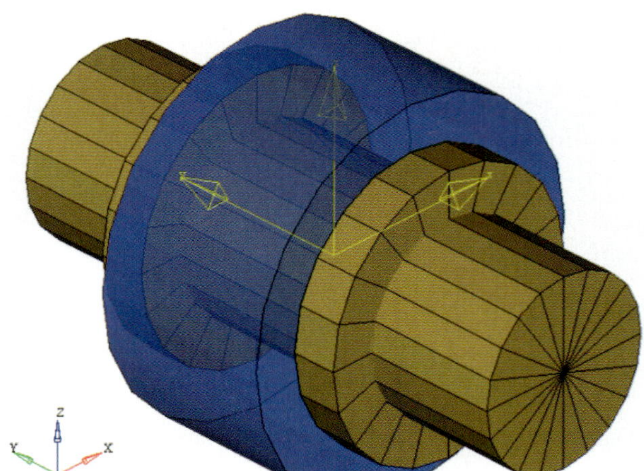

Fig. 2.8 Turning pair/revolute pair/hinge that allows only a rotary motion

Sliding Pair [Prismatic Pair]

Pair that allows only a rectilinear translation between two links, see Figure 2.9.

Screw Pair [Helical Pair]

Pair that allows only a screw motion between two links, see Figure 2.10.

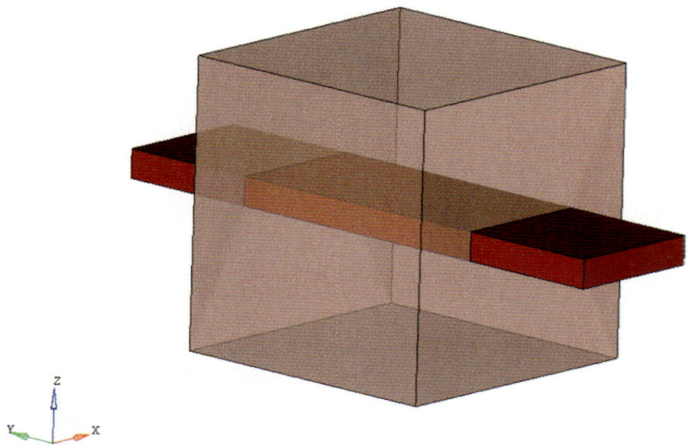

Fig. 2.9 Sliding pair/prismatic pair that allows only a rectilinear translation

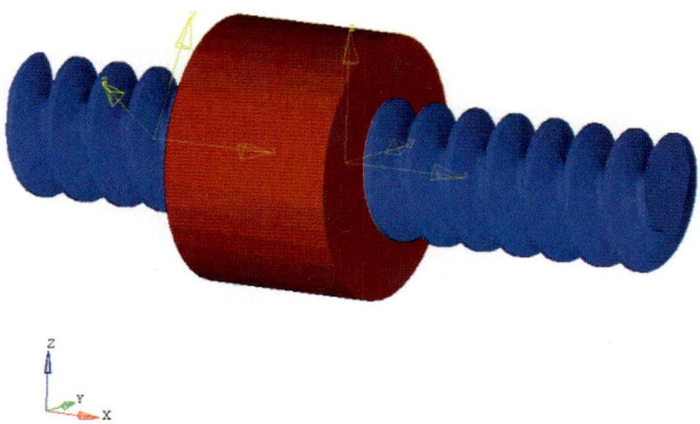

Fig. 2.10 Screw or helical pair

Lower Pair

Kinematic pair that is formed by surface contact between its elements. If A and B form a pair, the path traced by any point on element A relative to element B is identical with the path traced by any point on element B relative to element A.

Higher Pair

Kinematic pair that is formed by point or line contact between its elements.

Fig. 2.11 (a) Three links form a structure. (b) Simple kinematic chain with four links

2.2 Elementary Mechanisms

Minimum number of links to form a kinematic chain is three, as shown in Figure 2.11a. Since no relative motion is possible between these links, it forms a structure. The simplest kinematic chain has four links, as shown in Figure 2.11b.

Increasing number of links from four to five as in Figure 2.12a, the constraint of the system is completely lost. If the number of links is increased to six as in Figure 2.12b, the constraint is regained. So we see it is necessary to develop a relation that will tell us whether a given number of links can form a kinematic chain or not.

2.3 Grübler's Criterion for Planar Mechanisms

The number of degrees of freedom, F, of a planar mechanism with n links, j lower kinematic pairs and h higher kinematic pairs is

$$F = 3(n - 1) - 2j - h \tag{2.1}$$

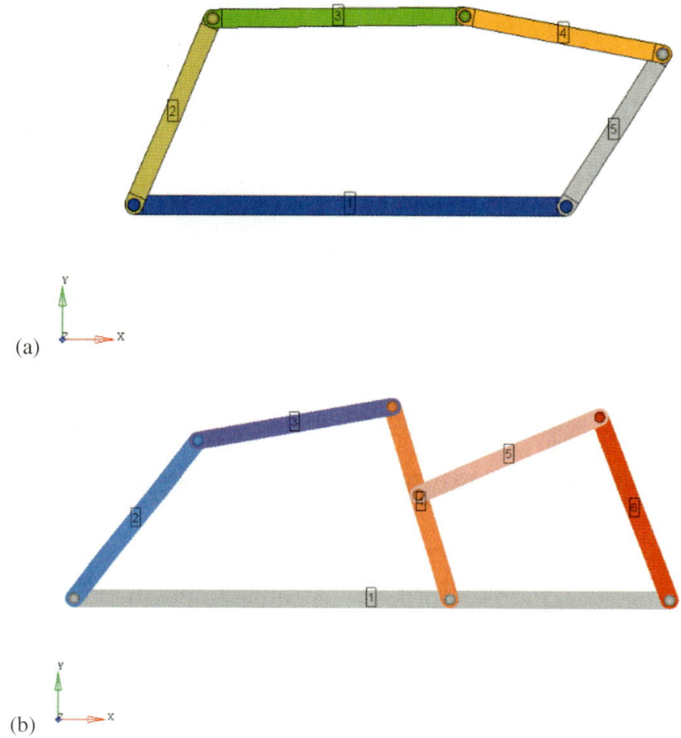

Fig. 2.12 (a) Five-link chain has no constrained motion. (b) Six-link chain has constrained motion

For constrained motion ($F = 1$)

$$2j - 3n + h + 4 = 0 \tag{2.2}$$

Although Grübler's criterion is applicable in almost all cases, a few exceptions exist, e.g., a fly-press shown in Figure 2.13.

Let us take an example of a six-bar linkage shown in Figure 2.14. Determine the degrees of freedom. There are four binary links and two ternary links. The number of joints are (you can count them directly or use the following formula)

$$\begin{aligned} j &= \frac{1}{2}(2n_2 + 3n_3) \\ &= \frac{1}{2}(2 \times 4 + 3 \times 2) = 7 \end{aligned}$$

The number of degrees of freedom from equation (2.1) is

$$F = 3(6 - 1) - 2 \times 7 = 1$$

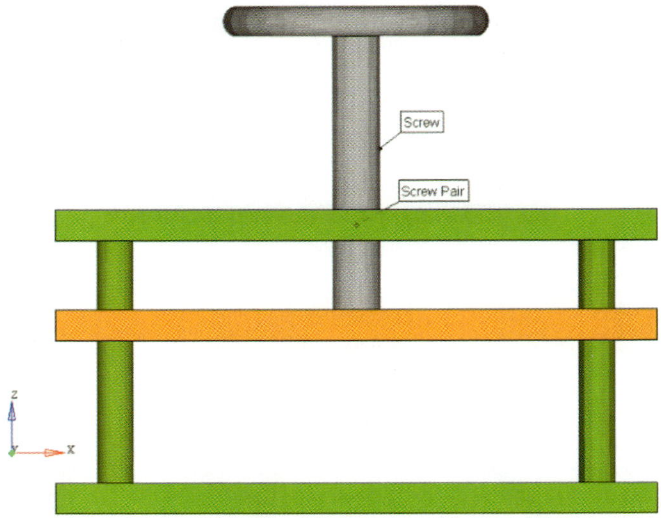

Fig. 2.13 A fly-press – an exception to Grübler's criterion

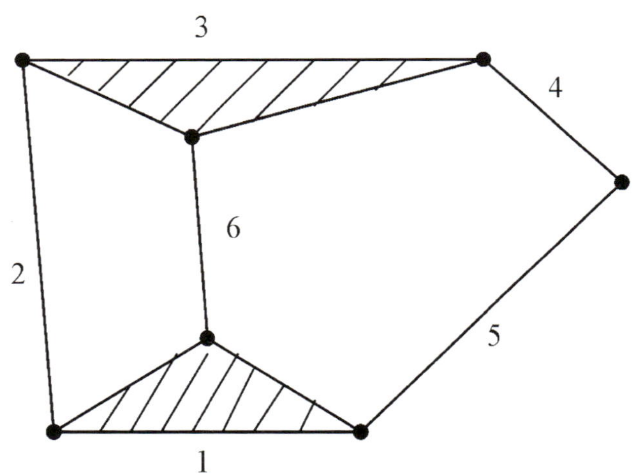

Fig. 2.14 A six-bar linkage

Thus, this linkage has one degree of freedom. If the link 1 is fixed to a frame and the link 2 is driven by a motor, the motions of the rest of the links 3 to 6 will be unique.

Consider another example, an eight-bar linkage shown in Figure 2.15. Determine the degrees of freedom.

There are five binary links ($n_2 = 5$), two ternary links ($n_3 = 2$) and one quaternary link ($n_4 = 1$). The number of joints is

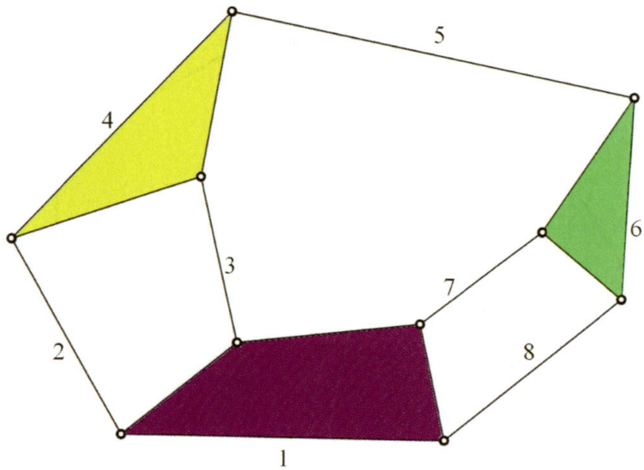

Fig. 2.15 Eight-bar linkage

$$j = \frac{1}{2}(2 \times 5 + 3 \times 2 + 4 \times 1) = 10$$

The number of degrees of freedom from equation (2.1) is

$$F = 3(8 - 1) - 2 \times 10 = 1$$

Thus, this linkage has also one degree of freedom. If the link 1 is fixed to a frame and the link 2 is driven by a motor, the motions of the rest of the links 3 to 8 will be unique.

Now let us consider a seven-bar linkage in Figure 2.16, the number of degrees of freedom can be shown to be two. If the link 1 is fixed to a frame, we need two inputs, e.g., links 2 and 5 to be driven, and then the motions of the rest of the links 3, 4, 6 and 7 will be unique.

Finally consider a six-bar linkage in Figure 2.17 with link 6 having a sliding motion on the fixed frame link 1. There are two binary links, 3 and 6; the remaining four are ternary links. Here, the fixed link 1 has two lower pairs (joints or hinges) and one higher pair (sliding pair) with link 6 which has a lower pair with link 5, therefore it is a ternary link. The number of joints (lower pairs) are $j = 7$ and there is one higher pair $h = 1$. Hence

$$F = 3(6 - 1) - 2 \times 7 - 1 = 0$$

Thus, this linkage has no degrees of freedom. It forms a structure.

Finally, let us show that the five-bar linkage in Figure 2.18 is not capable of producing relative motion. $N = 5$, $j = 6$ and therefore $F = 0$.

The study in Section 2.3 helps us in setting up a number of links with an appropriate number of elementary pairs, lower or higher to obtain a desired motion

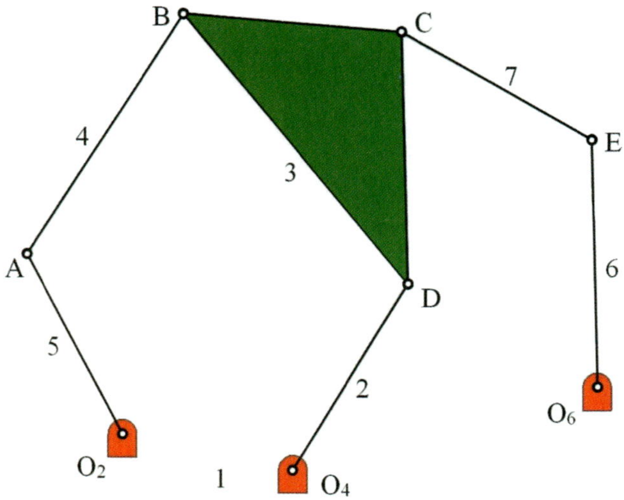

Fig. 2.16 Seven-bar linkage

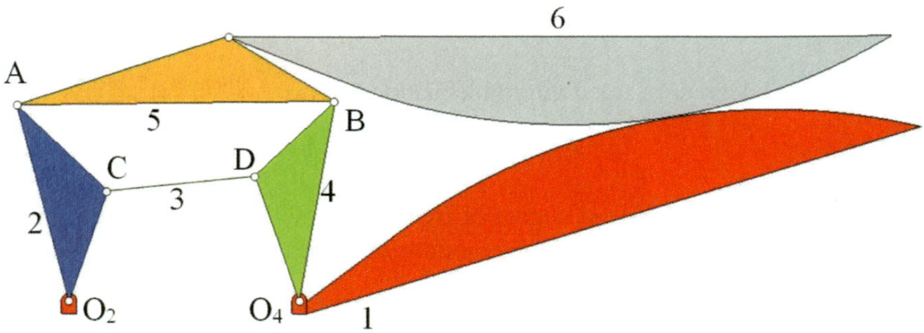

Fig. 2.17 Six-bar linkage

of a new machine or analyze an existing machine. Now let us consider a simple kinematic chain and see what we can do with that.

2.4 Four-Link Chains

Four-link chains with lower pairs are schematically represented in Figures 2.19a, 2.19b, 2.19c and 2.19d.

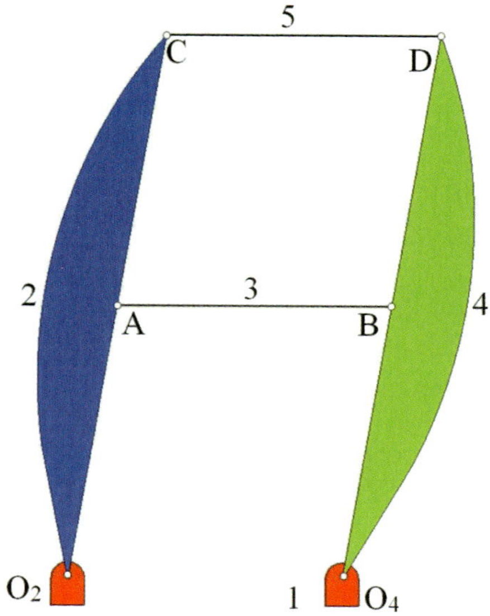

Fig. 2.18 Five-bar linkage

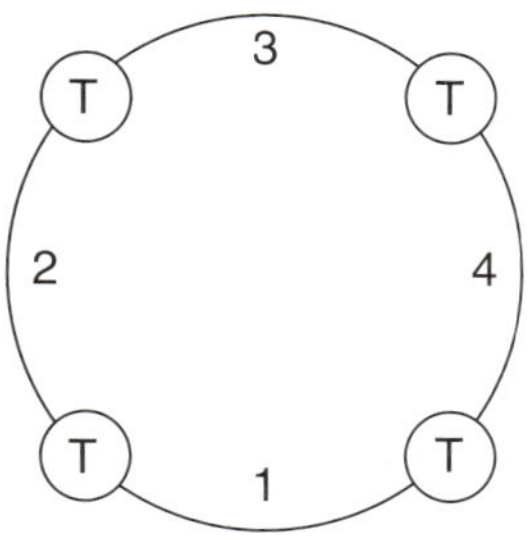

Fig. 2.19 **(a)** Quadric cycle chain with all turning pairs

Quadric Cycle Chain

Four-link chain with all turning pairs (joints or hinges), see Figure 2.19a.

Single Slider Chain

A quadric cycle chain with one of its turning pair replaced by a sliding pair, see Figure 2.19b.

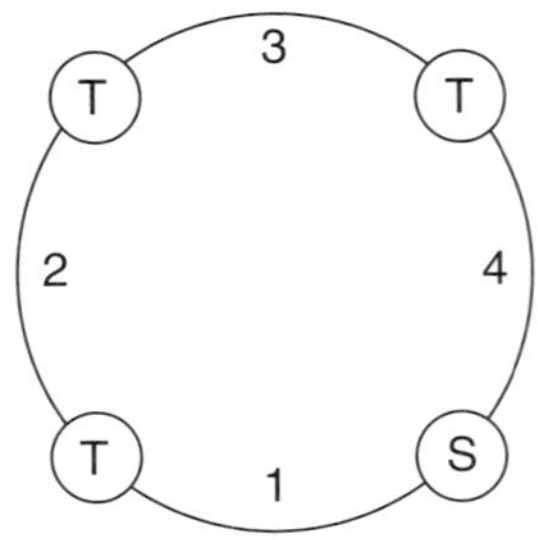

Fig. 2.19 (b) Quadric cycle chain with single slider

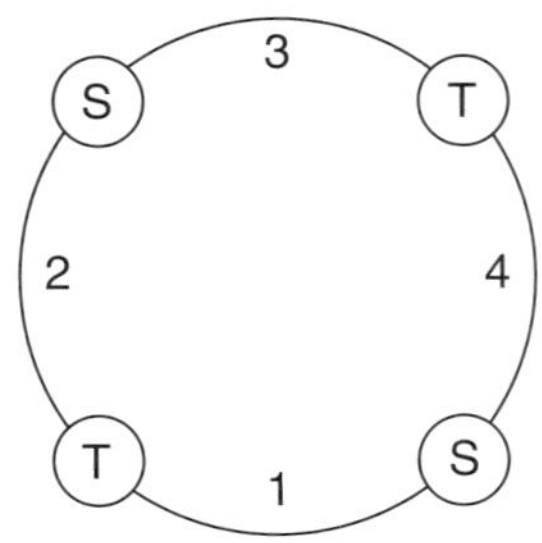

Fig. 2.19 (c) Crossed double slider quadric cycle chain

Crossed Double Slider Chain

A quadric cycle chain with two sliding pairs located opposite to each other, Figure 2.19c.

Double Slider Chain

A quadric cycle chain with two sliding pairs located next to each other, Figure 2.19d.

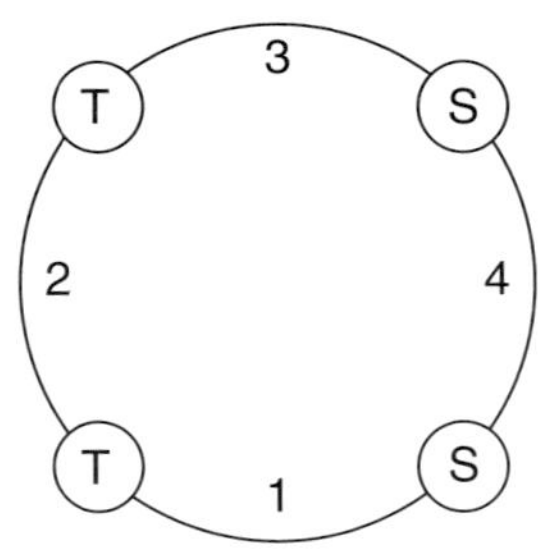

Fig. 2.19 (d) Double slider quadric cycle chain with adjacent sliding pairs

2.5 Kinematic Inversion

Four-Bar Linkage

Linkage with four binary links.

Four-Bar Mechanism

Mechanism with four binary links.

Crank

Link that rotates completely about a fixed axis.

Rocker [Lever]

Link that oscillates within a limited angle of rotation about a fixed axis.

Input [Driving] Link

Link where by motion and force are imparted to a mechanism.

Output [Driven] Link

Link from which required motion and forces are obtained.

Coupler [Floating] Link

Link that is not directly connected to the fixed link or frame.

Slider

Link that forms a prismatic pair (sliding pair) with one link and a revolute (turning) pair with another link.

Sliding Block

Compact element of a prismatic pair that slides along a guiding element.

Guide

Element of a prismatic pair that is fixed to a frame and constrains the motion of a sliding block.

Crosshead

Component between a piston and a connecting rod which, by forming a prismatic joint with the frame, provides a reaction to the component of force in the connecting rod normal to the line of stroke of the piston.

Connecting Rod

Coupler between a piston and or a cross-head and a crank shaft.

Kinematic Inversion

Transformation of one mechanism into another by choosing a different member to be the frame (fixed link).

We defined a mechanism in Section 2.1 as a kinematic chain with one of its components (link or joint) connected to the fixed frame. We can choose any one of the four links in a quadric cycle chain as a ground link that will produce different versions of mechanisms from the same quadric cycle chain. The mechanisms thus obtained are Kinematic Inversions of the original kinematic linkage.

Inversions of Quadric Cycle Chain

Kinematically speaking all inversions of a Quadric Cycle Chain are the same, however, by suitably altering the lengths of links, l_1, l_2, l_3 and l_4, different mechanisms can be obtained. These are described here.

1. *Crank-and-Rocker mechanism* (*Crank-Lever mechanism*): This mechanism is shown in Figure 2.20. Link 2 is the crank and link 4 is the rocker or lever. Link proportions for this case are:

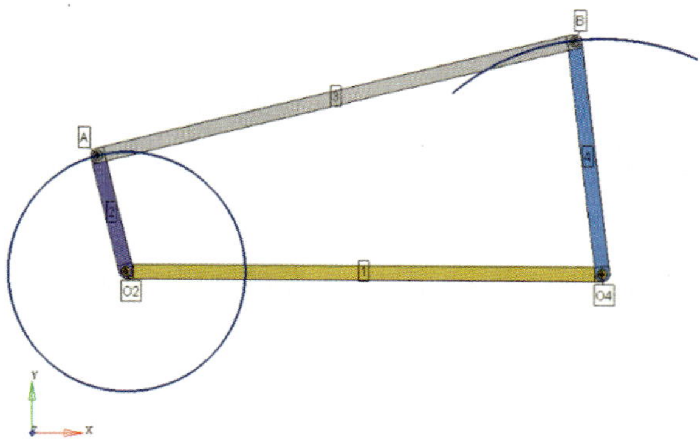

Fig. 2.20 Crank-lever mechanism

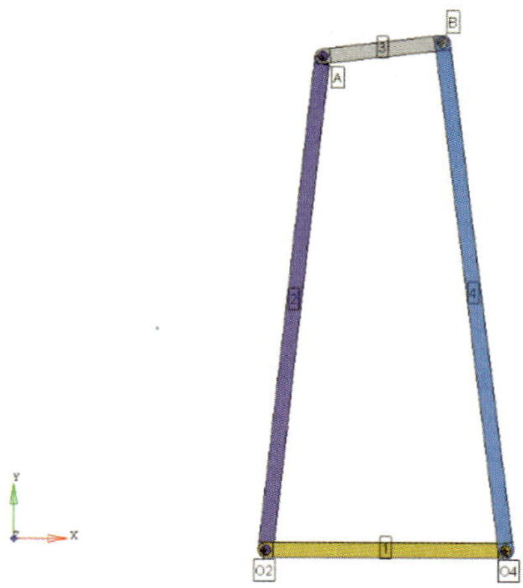

Fig. 2.21 Double-lever mechanism

$$(l_2 + l_3) < (l_1 + l_4)$$
$$(l_3 - l_2) > (l_1 - l_4) \tag{2.3}$$

2. *Double-Rocker mechanism* (*Double-Lever mechanism*): This is a four-bar mechanism with two rockers as shown in Figure 2.21. In this case

$$(l_3 + l_4) < (l_1 + l_2)$$
$$(l_2 + l_3) < (l_1 + l_4) \tag{2.4}$$

3. *Double-Crank mechanism*: Four-bar mechanism with two cranks.

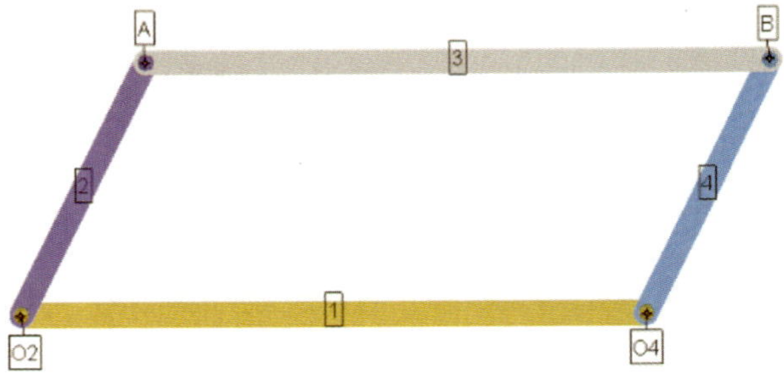

Fig. 2.22 Parallel-crank mechanism

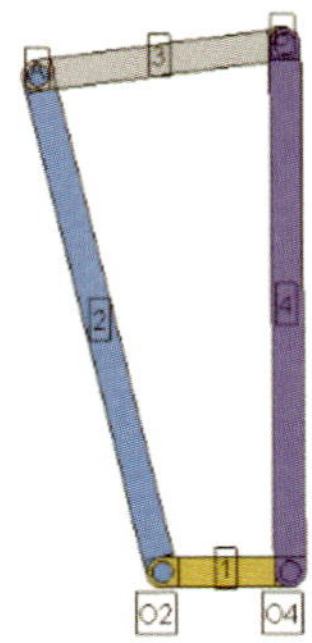

Fig. 2.23 Drag-link mechanism

a. *Parallel-Crank mechanism* Four-bar mechanism having cranks of equal length and a coupler with length equal to that of the fixed link (frame), see Figure 2.22.
b. *Drag-Link mechanism*, see Figure 2.23. In this case, the ground link is the shortest one and the coupler should be longer than that. Both the input and output links perform complete 360° revolutions. The coupler satisfies the following conditions:

$$
\begin{aligned}
& l_3 > l_1 \quad \text{and} \quad l_4 > l_2 \\
& l_3 > (l_1 + l_4 - l_2) \\
& l_3 < (l_2 + l_4 - l_1)
\end{aligned}
\tag{2.5}
$$

Branching Condition [Change Point Condition]

When

$$(l_1 + l_3) = (l_2 + l_4) \tag{2.6}$$

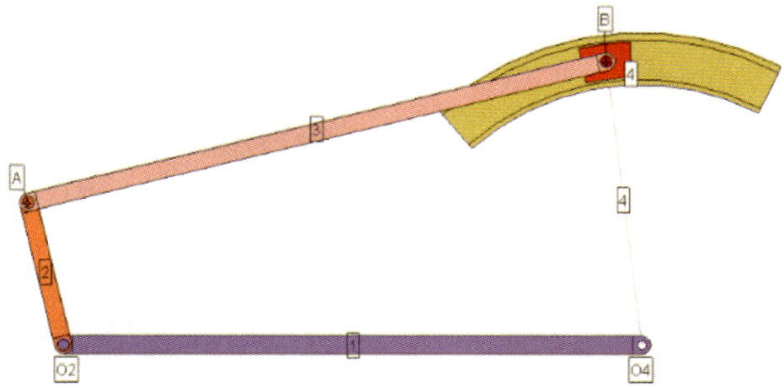

Fig. 2.24 Single slider chain obtained by replacing one revolute pair by a slider

the mechanism suffers from branching or a change point condition. At the change point, the center lines of all links become collinear and the output link may suffer a change in direction unless additional guidance is provided. (Note that a given four-bar mechanism can be drawn in two configurations, one normal-linked and the other cross-linked.)

Now let us find all the inversions of a quadric cycle chain with $l_1 = 10$, $l_2 = 20$, $l_3 = 30$ and $l_4 = 40$ cm.

1. *Crank-and-Rocker mechanism (Crank-Lever mechanism)*
   (a) $l_1 = 20$, $l_2 = 10$, $l_3 = 30$ and $l_4 = 40$ cm (Change Point)
   (b) $l_1 = 20$, $l_2 = 10$, $l_3 = 40$ and $l_4 = 30$ cm
   (c) $l_1 = 30$, $l_2 = 10$, $l_3 = 20$ and $l_4 = 40$ cm (Change Point)
   (d) $l_1 = 30$, $l_2 = 10$, $l_3 = 40$ and $l_4 = 20$ cm
   (e) $l_1 = 40$, $l_2 = 10$, $l_3 = 20$ and $l_4 = 30$ cm
   (f) $l_1 = 40$, $l_2 = 10$, $l_3 = 30$ and $l_4 = 20$ cm

2. *Double-Rocker mechanism (Double-Lever mechanism)*
   (a) $l_1 = 40$, $l_2 = 20$, $l_3 = 10$ and $l_4 = 30$ cm
   (b) $l_1 = 30$, $l_2 = 20$, $l_3 = 10$ and $l_4 = 40$ cm
   (c) $l_1 = 20$, $l_2 = 30$, $l_3 = 10$ and $l_4 = 40$ cm

3. *Double-Crank mechanism*
   (a) $l_1 = 10$, $l_2 = 20$, $l_3 = 30$ and $l_4 = 40$ cm
   (b) $l_1 = 10$, $l_2 = 20$, $l_3 = 40$ and $l_4 = 30$ cm (Change Point)
   (c) $l_1 = 10$, $l_2 = 30$, $l_3 = 20$ and $l_4 = 40$ cm

Inversions of Single Slider Chain

Link 4 of a four-bar mechanism with turning pairs is replaced by a slider; see Figure 2.24 to form a single slider chain.

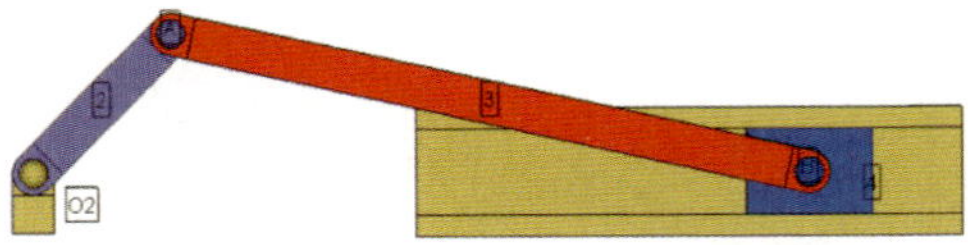

Fig. 2.25 Reciprocating engine mechanism

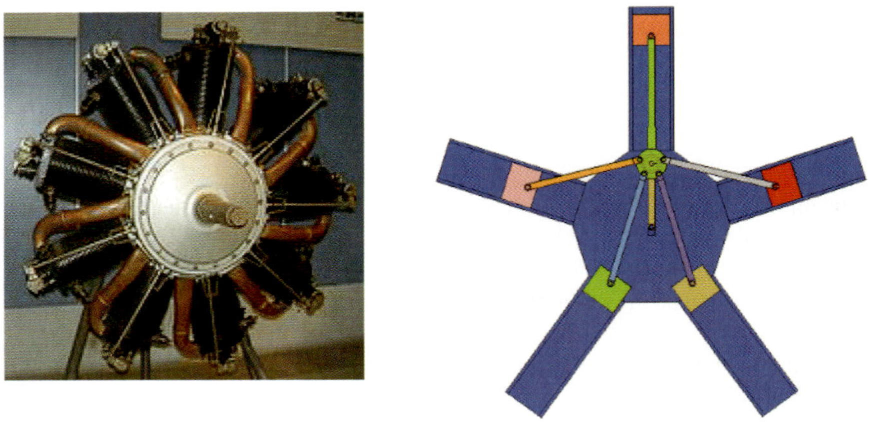

Fig. 2.26 Rotary engine with fixed crank

1. *Inversion with link 1 fixed*: Reciprocating engine mechanism, see Figure 2.25. This is the most common mechanism used to day in all internal combustion engines.
2. *Inversions with link 2 fixed*:

   a. *Rotary Engine*: See Figure 2.26. The engine shown has five cylinders. Out of the five connecting rods, one is a master connecting rod and the other four are slave rods. The crank 2 is common to all five cylinders and is fixed. Link 1 is the engine block which rotates. Le Rhône 9C, a typical rotary engine of WWI is shown on the left. The copper pipes carry the fuel-air mixture from the crankcase to the cylinder heads.
   The design was used mostly in the years shortly before and during World War I to power aircraft, and also saw use in a few early motorcycles and cars. By the early 1920s the rotary aircraft engine was becoming obsolete, mainly because of an upper ceiling to its possible output torque, which was a fundamental consequence of the way the engine worked. It was also limited by its inherent restriction on breathing capacity, due to the need for the fuel/air mixture to be aspirated through the hollow crankshaft and crankcase, which directly affected its volumetric efficiency. However, at the time it was a very efficient solution to the problems of power output, weight, and reliability. The

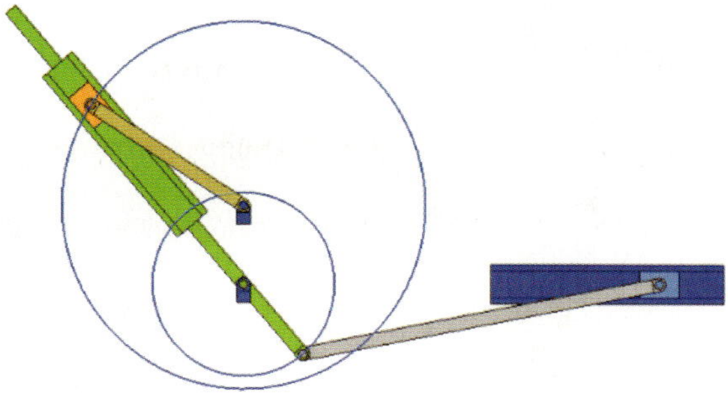

Fig. 2.27 Whitworth quick return motion mechanism

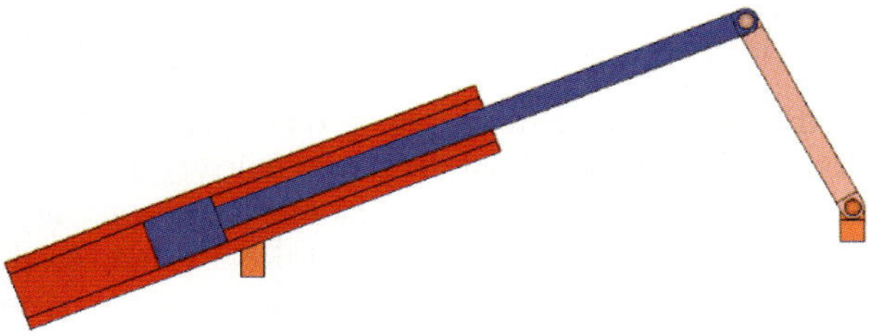

Fig. 2.28 Oscillating cylinder engine

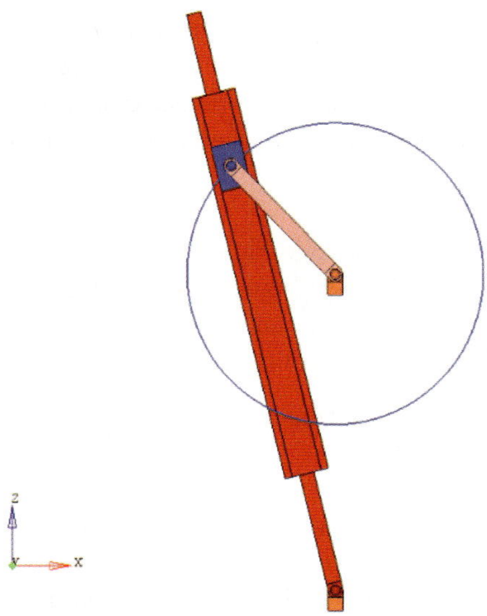

Fig. 2.29 Quick return mechanism

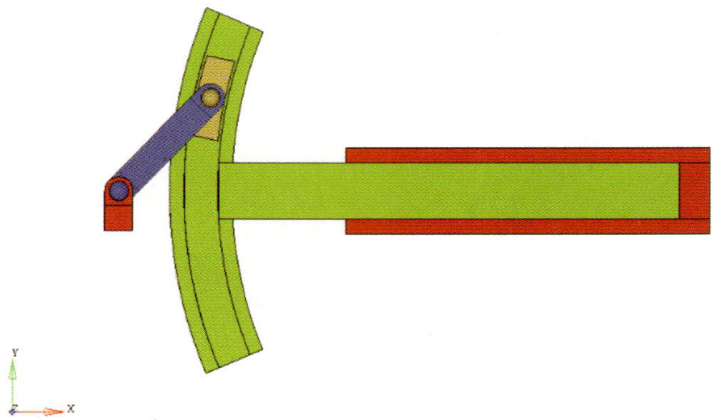

Fig. 2.30 Formation of a double-slider chain

rotation of the bulk of the engine's mass produced a powerful gyroscopic flywheel effect, which smoothed out the power delivery and reduced vibration. Vibration had been such a serious problem on conventional piston engine designs that heavy flywheels had to be added. Rotary and radial engines look strikingly similar when they are not running and can easily be confused, since both have cylinders arranged radially around a central crankshaft.

b. *Whitworth Quick Return mechanism*: See Figure 2.27. Link 3 is the crank here. Slider 4 drives link 1. Link 1 drives a cutting tool through the connecting rod 5. The forward stroke starts with link 3 in position *AQ* and ends at *AP* through *AS*. The return motion is faster from *AP* through *AR* to *AQ*. These linkages are most useful in saving time, since the return stroke, which is an idle stroke, is faster than a useful forward stroke when metal is removed.

3. *Inversions with link 3 fixed*

   a. *Oscillating Cylinder Engine*: See Figure 2.28.
   b. *Quick Return Mechanism*: See Figure 2.29. Link 2 is the crank; link 4 is the rocker arm. The forward stroke starts with link 2 in position *P* and ends at *Q* through *S*. The return motion is faster from *Q* through *R* to *P*.

4. *Inversion with link 4 fixed*: These are same as case 1.

Inversions of Double Slider Chain

Figure 2.30 illustrates how a double slider chain is formed.

1. *Inversion with link 1 fixed*: Scotch–Yoke mechanism, see Figure 2.31. This is a four-bar mechanism in which a crank is connected by a slider with another link which, in turn, forms a prismatic pair with the frame. Kinematically, this is the same as fixing link 3.

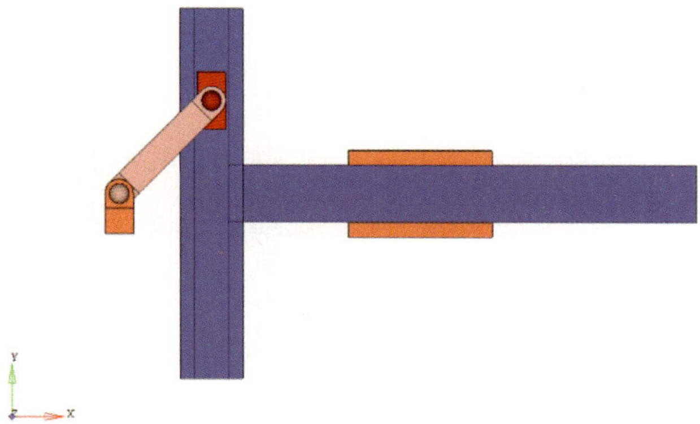

Fig. 2.31 Scotch–Yoke mechanism

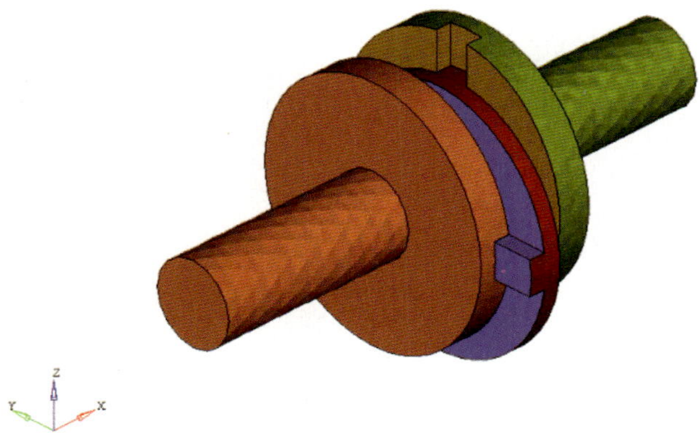

Fig. 2.32 Oldham's coupling

2. *Inversion with link 2 fixed*: This gives rise to Oldhams coupling in Figure 2.32. Links 1 and 3 have slotted grooves to form sliding pairs with corresponding faces of link 4.
3. *Inversion with Link 4 fixed*: This gives an Elliptic Trammel mechanism, see Figure 2.33.

We looked at a quadric cycle chain and various useful mechanisms it can generate and that are exploited in developing machines. In a similar way, one can explore multi-link chains and make a systematic study to identify possible new machines. This is out of the scope of the present book.

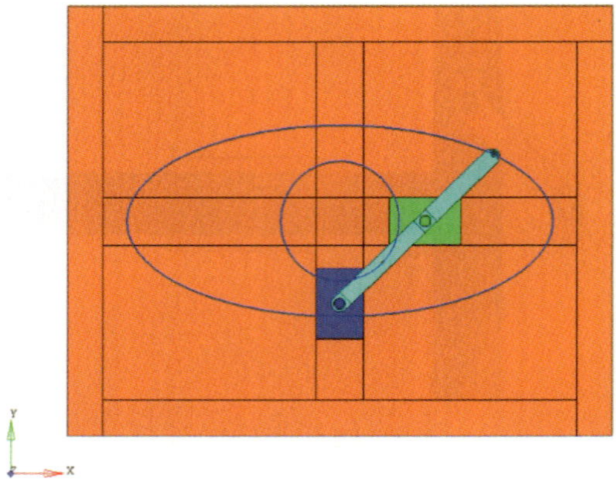

Fig. 2.33 Elliptic trammel mechanism

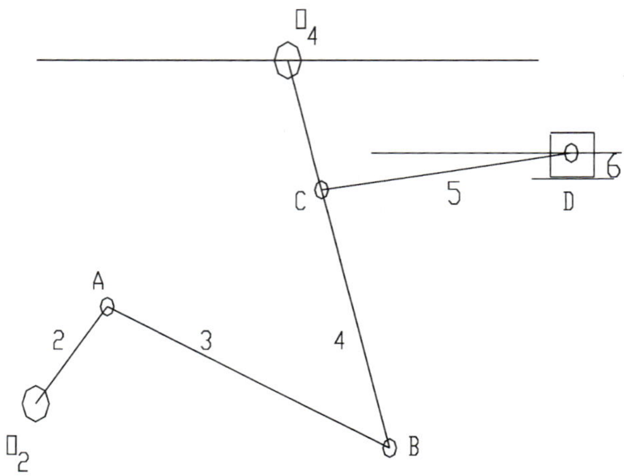

Fig. 2.34

2.6 Additional Problems

1. Refer to Figure 2.34. Identify (a) the number and type of links, (b) the different elements, (c) the kinematic pairs and their type, (d) draw a schematic diagram representing its kinematic chain, (e) is there a basic quadric cycle chain in the linkage, if so what type, and (f) determine the number of degrees of freedom.
2. Refer to Figure 2.35. Identify (a) the number and type of links, (b) the different elements, (c) the kinematic pairs and their type, (d) draw a schematic diagram representing its kinematic chain, (e) is there a basic quadric cycle chain in the linkage, if so what type, and (f) determine the number of degrees of freedom.
3. Refer to Figure 2.36. Identify (a) the number and type of links, (b) the different elements, (c) the kinematic pairs and their type, (d) draw a schematic diagram

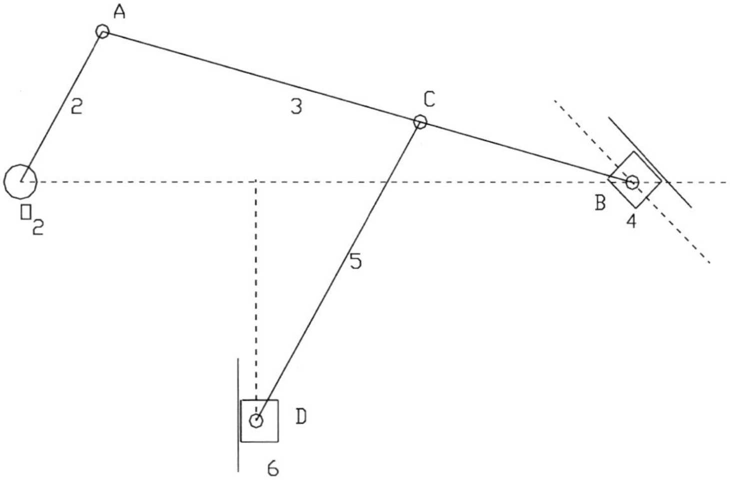

Fig. 2.35

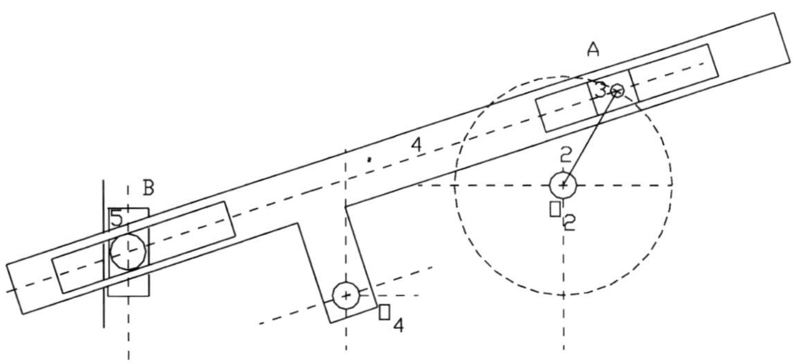

Fig. 2.36

representing its kinematic chain, (e) is there a basic quadric cycle chain in the linkage, if so what type, and (f) comment on the number of degrees of freedom.

4. Refer to Figure 2.37. Compare this linkage with that of Figure 2.34 above and make any comments you have.
5. Refer to Figure 2.38. Identify (a) the number and type of links, (b) different elements, (c) kinematic pairs and their type, (d) draw a schematic diagram representing its kinematic chain, (e) is there a basic quadric cycle chain in the linkage, if so what type, and (f) determine number of degrees of freedom.

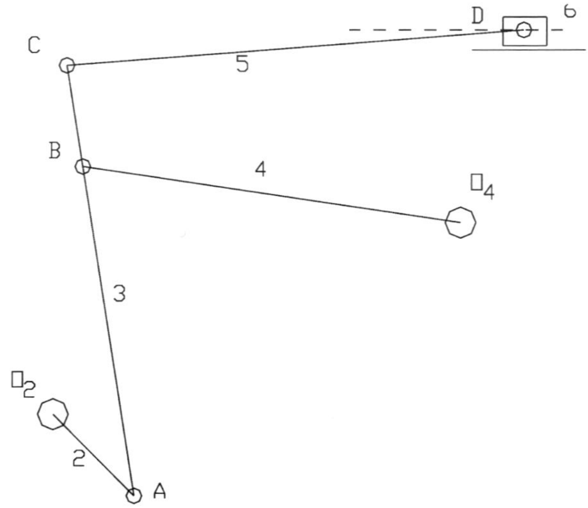

Fig. 2.37

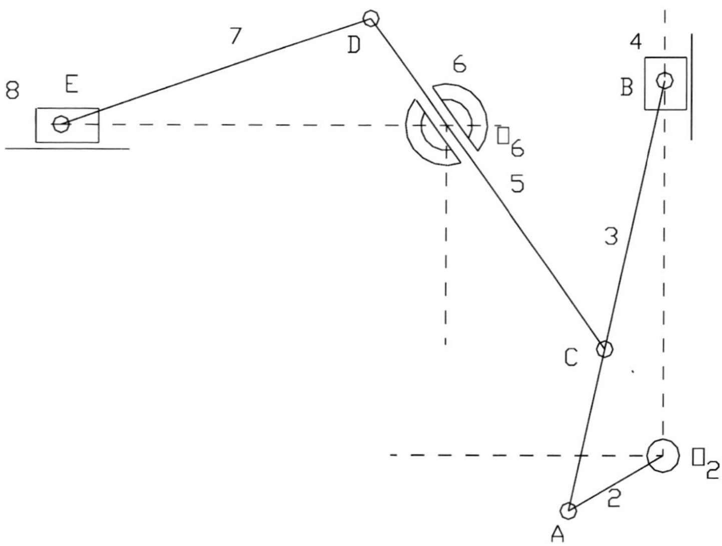

Fig. 2.38

Chapter 3
Kinematic Analysis of Mechanisms

Kinematics

Branch of theoretical mechanics dealing with the geometry of motion, irrespective of the causes that produce the motion.

Kinematic Analysis

Analysis of the kinematic aspects of mechanisms.

Motion

Changing position of a body relative to a frame of reference.

Absolute Motion

Motion with respect to a fixed frame of reference.

Relative Motion

Motion with respect to a moving frame of reference.

Translation

Motion of a rigid body in which each straight line rigidly connected with the body remains parallel to its initial direction.

Rectilinear Translation

Translation in which the paths of points of a rigid body are straight lines. This is linear motion, e.g., the piston in a reciprocating engine mechanism.

Rotation

Motion of a rigid body in which all its points move on circular arcs centered on the same axis, e.g., the crank in a reciprocating engine mechanism.

Plane [Planar] Motion

Motion of a rigid body in which its points describe curves located in parallel planes. This is a combination of translation and rotation; see Figure 3.1, e.g., the connecting-rod in a reciprocating engine mechanism.

Displacement

Change of position of a body with respect to a fixed frame of reference.

Relative Displacement

Displacement with respect to a moving frame of reference.

Angular Displacement

Displacement of a rigid body in rotation.

Velocity

Rate of displacement with respect to time.

Absolute Velocity

Velocity with respect to a fixed frame of reference.

Relative Velocity

Velocity with respect to a moving frame of reference.

Instantaneous Center [of Velocity]

Point in a rigid lamina moving in its own plane where the velocity relative to a frame of reference is zero at the given instant; see Figure 3.1.

Acceleration

Rate of change of velocity with respect to time.

Normal Acceleration

Component of acceleration of a point normal to its velocity.

Tangential Acceleration

Component of acceleration of a point collinear with its velocity.

Absolute Acceleration

Rate of change of absolute velocity with respect to time.

Relative Acceleration

Rate of change of relative velocity with respect to time.

Centripetal Acceleration

Acceleration of a point towards the center of curvature of its path as it moves along a fixed curve.

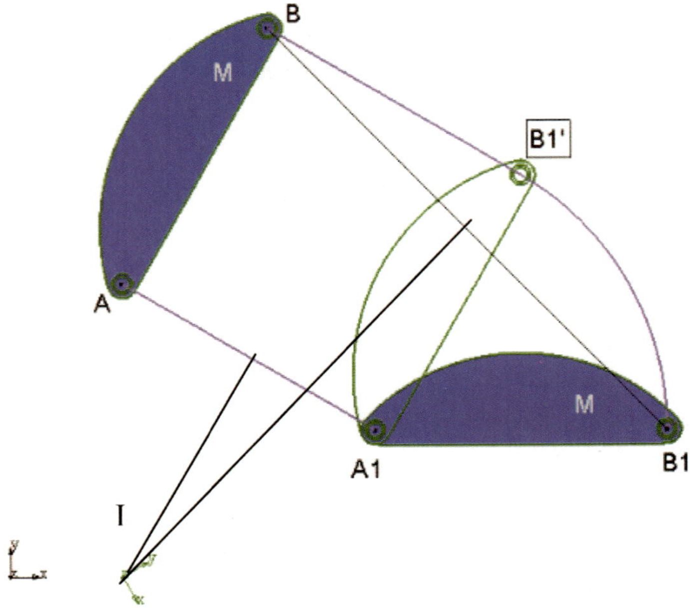

Fig. 3.1 General motion of a rigid body M

Coriolis Acceleration

Component of the absolute acceleration of a point due to its velocity relative to a rotating frame of reference. It equals twice the vector product of the angular velocity of the moving frame of reference and the relative velocity of the given moving point.

Angular Acceleration

Rate of change of angular velocity with respect to time.

3.1 Velocities by the Centro Method

Of all the motions, rectilinear translation and rotation are simple compared to general plane motion or space motion. For a general motion we can always imagine a body to be rotating about a center instantaneously. The center of rotation will be changing from instant to instant for a general motion. Therefore the concept of instantaneous center for rotation becomes handy.

However in a linkage the motion of each link is dependent on others as well and we need to develop a method for obtaining the instantaneous centers of rotation of all links for different linkage configurations. This method that depends on the concept of rotation is the Centro Method.

Centro

Two coincident points, belonging to two rigid bodies having relative motion with the following properties:

1. they have the same velocities; and
2. they form a point in one of the rigid bodies about which the other rotates and vice-versa, which is perhaps true for only an instant.

Primary Centro

One that can be easily located by a mere observation of the mechanism, see Figure 3.2.

Secondary Centro

Centro that cannot be easily located.

Kennedy's Theorem

For any three rigid bodies having relative motion, there are three centros all of which lie on a straight line.

Proof. Consider three rigid bodies 1, 2 and 3 having relative motion as shown in Figure 3.3. Link 1 is the reference link (frame). There are three centros, two primary centros 12, 13 and a secondary Centro 23.

Let any arbitrary point P represent the Centro 23. Considering P to be a part of body 2, its velocity V_{P2} will be as shown in Figure 3.3. Similarly, when P is considered as a part of body 3, V_{P3} will be its velocity. These two velocities should be identical, if centro P belongs to both bodies 2 and 3. This is possible only when P lies on the line joining the centros 12 and 13. This proves Kennedy's theorem.

Number of Centros in a Mechanism

For a mechanism of n links, the number of centros N is

$$N = \frac{1}{2} n (n - 1) \tag{3.1}$$

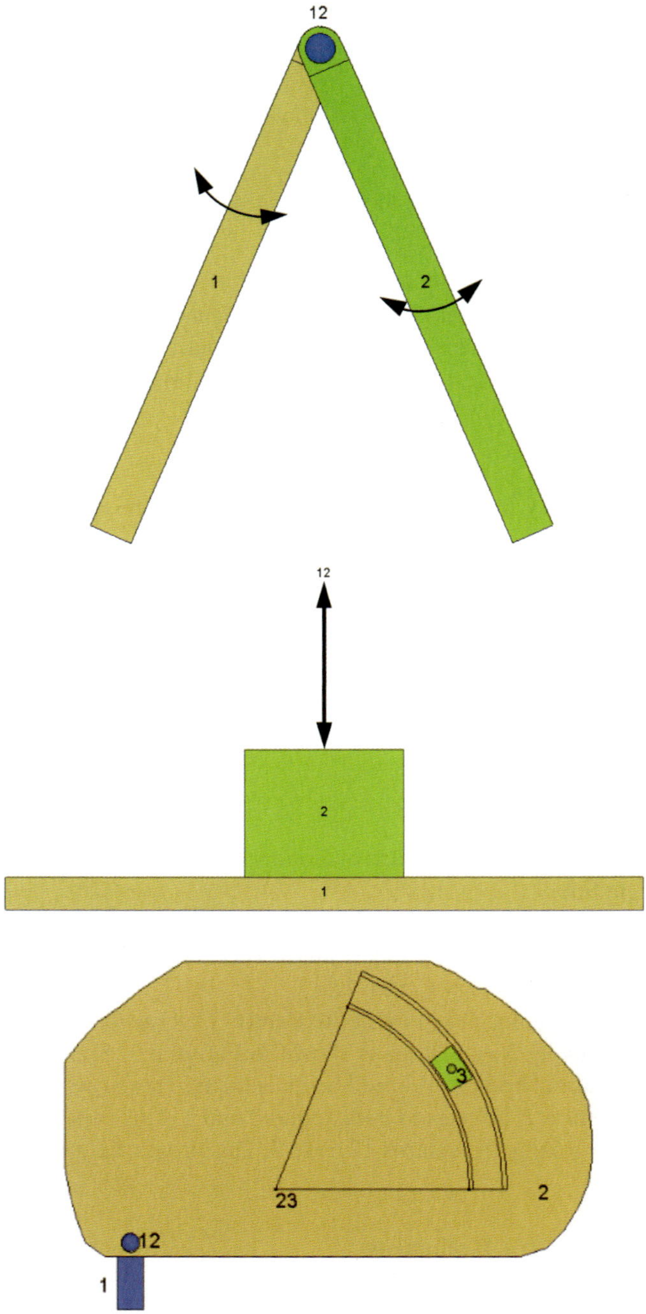

Fig. 3.2 Examples for locating primary centro

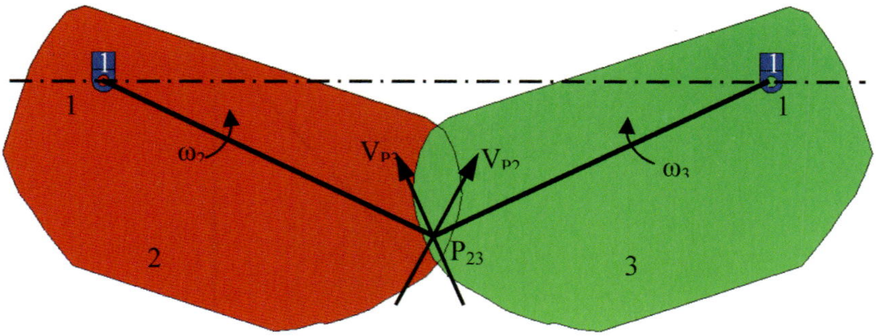

Fig. 3.3 Relative motion of three rigid bodies 1, 2 and 3

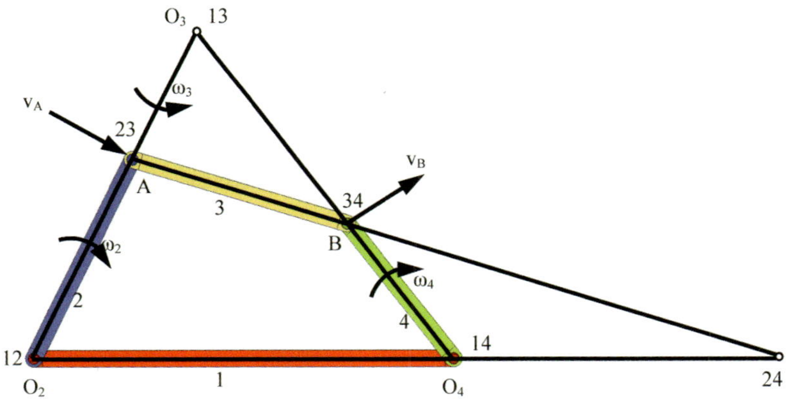

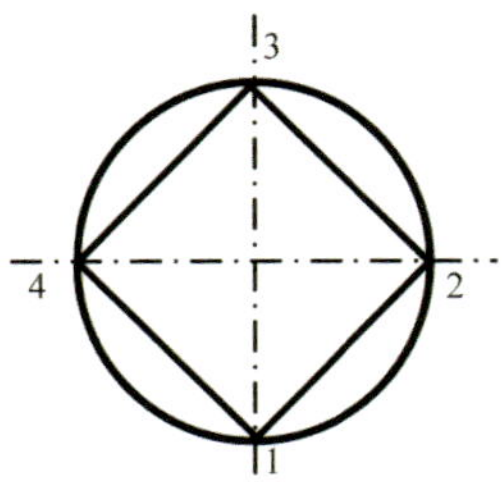

Table of Centros			
1	2	3	4
	~~12~~	13	~~14~~
		~~23~~	24
			~~34~~

Fig. 3.4 Determination of centros for a four-bar mechanism

Number of Lines of Centros

The number of lines of centros, L, for a mechanism with n links is

$$L = \frac{1}{6} n\,(n-1)\,(n-2) \tag{3.2}$$

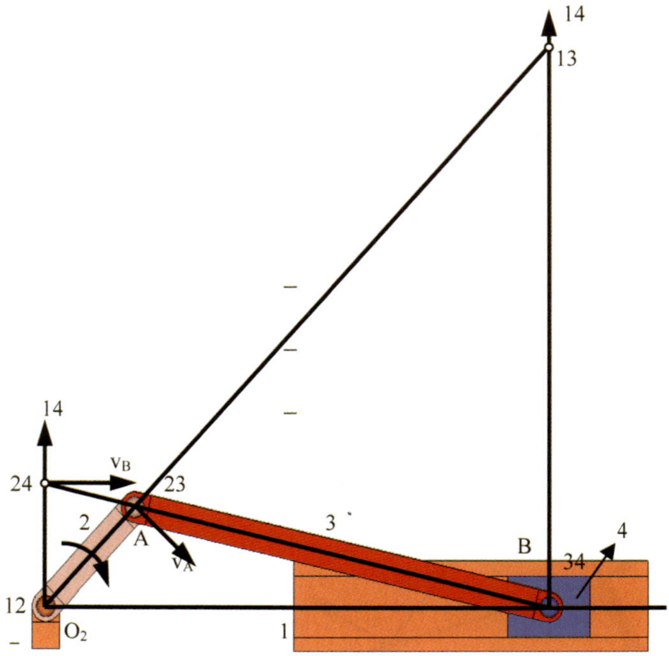

Fig. 3.5 Determination of centros for a slider-crank mechanism

Procedure to Determine Centros

The procedure to determine centros is described briefly for three different linkages.

1. *Four-Bar Linkage*
   Figure 3.4 gives a four-bar mechanism. Centros 12, 23, 34 and 14 are primary centros. To keep track of centros determined, use the circle of centros or table of centros. Primary centros are entered as full lines and secondary centros by dashed lines.
   To determine the secondary centros, use Kennedy's theorem repeatedly. To determine centro 23, choose two sets of three bodies, each set containing bodies 1 and 3 and any other body. For example, 1, 3 & 2 and 1, 3 & 4. Make sure that you have already located two centros out of three in these two sets (the unknown centro should form a diagonal in the circle). Centro 13 lies on 12–23 as well as 14–34. Similarly, centro 24 lies on 23–34 and 12–14.
2. *Slider Crank Chain*
   Figure 3.5 gives a slider-crank mechanism. As before, centros 12, 23, 34 and 14 are primary centros. Note that primary centro 14 is at infinity perpendicular to the path of the piston link 4.
3. *Eight-Link Chain*
   An eight-link mechanism is given in Figure 3.6. Secondary centros 13 and 15 are

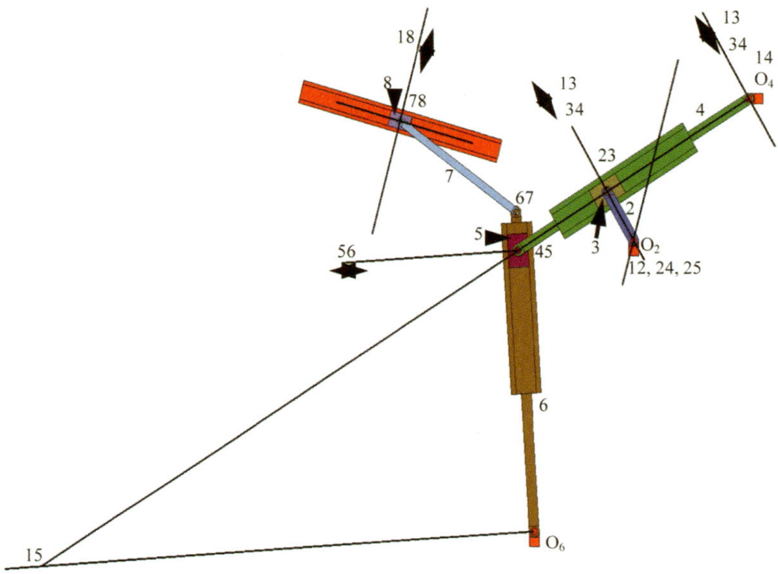

Fig. 3.6 Determination of centros for an eight-link mechanism

only located to illustrate the procedure. Determine the remaining centros as an exercise.

Determination of Velocities with the Help of Centros

Recall the following two properties:

- Every link in a mechanism has a common centro with the frame (fixed link). Therefore, every link can be treated as rotating about a fixed axis for an instant when the links of the mechanism occupy a given configuration.
- Every link has a common centro with the remaining links of the mechanism. That is, for every combination of two links, there is a common point belonging to both the links.

Two different methods can be used as described below.

(a) *Link to Link Method*. The following steps may be used.
Identify a known link, the link for which the velocities are known, e.g., the crank or the driving link (let this be link 2).

Determine the velocity of the centro belonging to the known link and the next link (i.e., centro 23). This centro is called a Transfer Centro, which enables the transfer of information from one link to another.

Now, the velocity of one point on link 3 is known. Its instantaneous center, i.e. centro with the ground link is also known. Therefore, the velocities of link 3 can be determined.

Identify the common centro of link 3 and its adjacent link 4 and determine its velocity.

Repeat the process until all the desired velocities are obtained.

This method will be useful, if the velocities of all the links in a mechanism are desired.

As an example consider the four-bar mechanism in Figure 3.4. Let crank 2 rotate with an angular velocity ω_2 rad/s. We follow the above mentioned steps.

1. Link 2 is the known link.
2. Centro 23 is at A. Its velocity is $V_A = \omega_2 \times O_2A(O_2A$ is measured in meters). We call the centro 23 a Transfer Centro, as it is used to transfer the information from one link to the next link.
3. O_3 (13) is the instantaneous center for link 3. Therefore, the angular velocity of link 3 is $\omega_3 = V_A/O_2A$ rad/s. Verify that this angular velocity is counter-clockwise (ccw).
4. The common centro between links 3 and 4 is at B. Its velocity in body 3 is $V_B = \omega_3 \times O_4B$ m/s.
5. O_4 (14) is the instantaneous center for link 4. Therefore, the angular velocity of link 4 is $\omega_4 = V_B/O_4B$ rad/s. Verify that this angular velocity is clockwise (cw) in direction.

(b) *Line of Centers Method*. In this method three links are chosen for the analysis, (1) known link K, (2) unknown link U and (3) frame or fixed link F. The following steps may be used:

1. Identify the three centros, the instantaneous centers KF, UF and the transfer centro KU. Since these centros belong to three bodies having relative motion, they should lie on one line, therefore the name Line of Centers method.
2. Determine the velocity of the centro KU.
3. Now, the velocity of one point on link U is known. Its instantaneous center, i.e., Centro with the fixed link UF, is also known. Therefore, the velocities of link U can be determined.

This method will be useful if the velocities of only a few links in a mechanism are desired.

As an example consider the reciprocating engine mechanism in Figure 3.5. Let crank 2 rotate with an angular velocity ω_2 rad/s. The velocity of the piston is desired. Here, the known link is 2, the fixed link is 1 and the piston is link 4.

1. Identify the three centros instantaneous centers 12, 14, and transfer centro 24.
2. The velocity of the transfer centro 24 is $V_{24} = \omega_2(12 - 24)$ m/s (measure the distances in meters).
3. Link 4 has a pure translation, therefore, the velocity of piston is $V_B = V_{24}$.

In another example of a Hack Saw mechanism of Figure 3.6, determine the velocity of slider 5, when crank 2 is rotating with an angular velocity ω_2 rad/s cw (clockwise).

Here, the known link is 2, the fixed link is 1 and the unknown link is 5.

1. The three centros required are, instantaneous centers 12, 15, and transfer centro 25.
2. The velocity of the transfer centro 25 is zero, since it coincides with the instantaneous center 12 (verify this yourself) in Figure 3.6.
3. Therefore, the velocity of link 5 is also zero.
4. From Figure 3.6, we note that crank 2 is connected to slider 3 through a hinge which is perpendicular to rocker 4. Therefore, the rocker at this instant is stationary. Hence, link 5 attached to the rocker through a hinge will also be stationary.
5. The Method of Centros is powerful for determining velocities in a linkage; however for determining accelerations in the next step, it becomes complicated. Hence we look for an alternate method; here we will discuss the Relative Velocity Equation method.

3.2 Relative Velocity Equation

The relative velocity equation approach to determine velocities is given here.

3.2.1 Rotation of a Rigid Link about a Fixed Axis

Consider the body M in Figure 3.7 rotating about a fixed axis O. The coordinates of point B in the OXY axis system are

$$x_B = R\cos\theta$$
$$y_B = R\sin\theta \tag{3.3}$$

θ is measured in the counter-clockwise direction. The velocities are obtained by differentiating the above with respect to time t. Let the angular velocity be $\omega = d\theta/dt$. Then

$$V_B^x = \frac{dx_B}{dt} = -R\omega\sin\theta$$
$$V_B^y = \frac{dy_B}{dt} = R\omega\cos\theta \tag{3.4}$$

The total velocity is given by

$$V_B = V_B^x + \rightarrow V_B^y = -R\omega\sin\theta + \rightarrow R\omega\cos\theta = R\omega \tag{3.5}$$

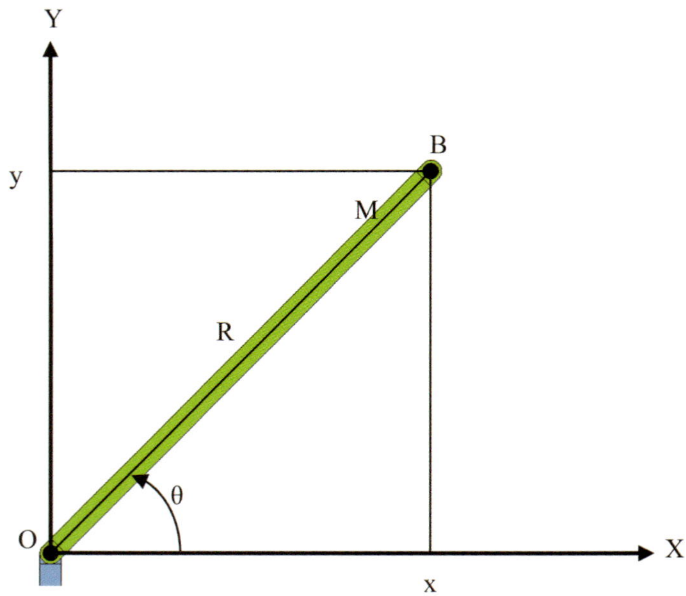

Fig. 3.7 Rotation of a rigid link about a fixed axis

This is shown vectorially in Figure 3.8. Since O is a fixed point, we normally do not make it a point to say that the velocity of point B is with respect to O. In general we can say that $V_{BO} = R\omega$.

3.2.2 Relative Velocity Equation of Two Points on a Rigid Body

Consider the body M to describe a general plane motion, see Figure 3.9. A is the pole, i.e., its velocity is known. We wish to find the velocity of B at a distance R making an angle θ with the positive direction of the X axis. The coordinates of point B are

$$x_B = x_A + R\cos\theta$$
$$y_B = y_A + R\sin\theta \tag{3.6}$$

The corresponding velocities are

$$V_B^x = V_A^x - R\omega\sin\theta$$
$$V_B^y = V_A^y + R\omega\cos\theta \tag{3.7}$$

The total velocity of B is

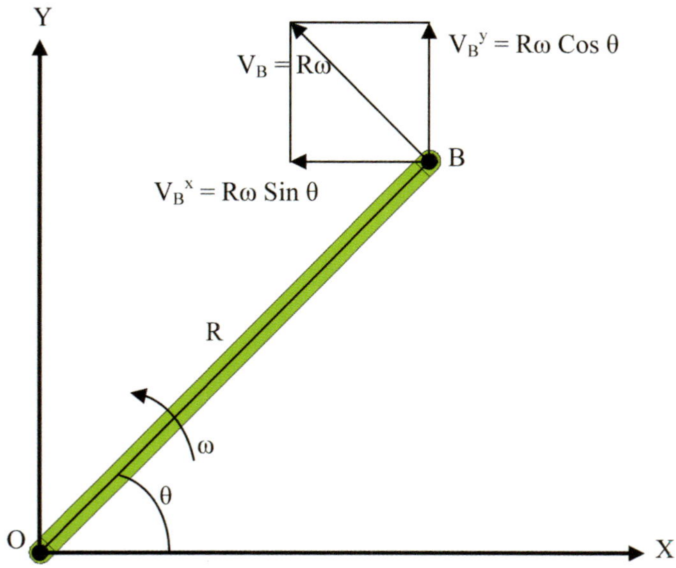

Fig. 3.8 Vectorial addition in equation (3.5)

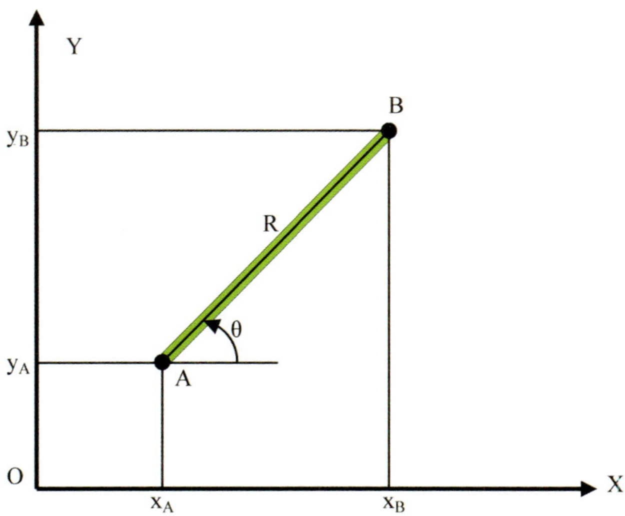

Fig. 3.9 A rigid body describing general motion

$$V_B = V_A^x + \rightarrow V_A^y + \rightarrow R\omega \sin\theta + \rightarrow R\omega \cos\theta = V_A + \rightarrow R\omega \qquad (3.8)$$

Vectorial addition of these components is shown in Figure 3.10. The total velocity of B consists of two parts: (1) translational component V_A, and (2) rotational component $R\omega$. These two components are shown in Figure 3.11. The rotational

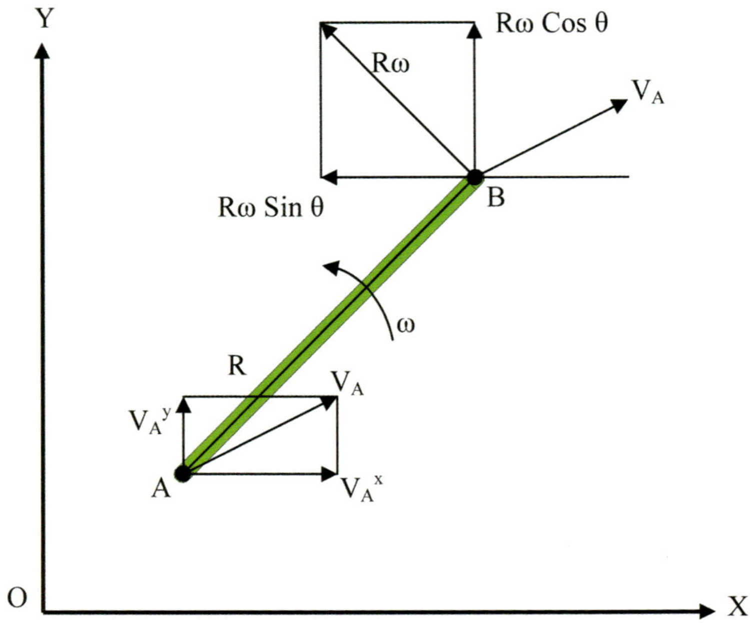

Fig. 3.10 Vector representation of total velocity of B

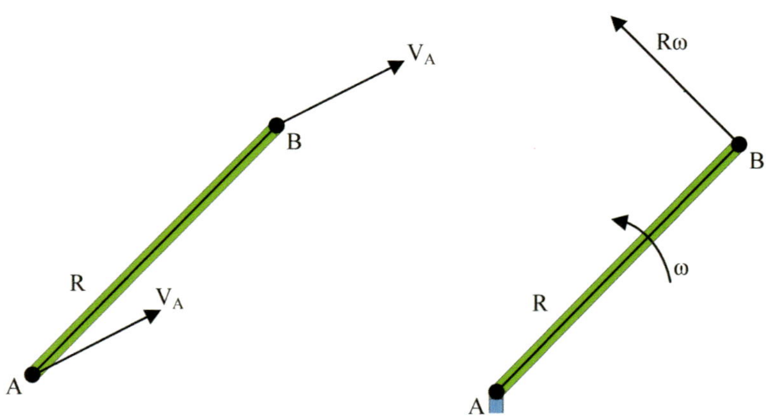

Fig. 3.11 Translational and rotational components of velocity of point B

component $R\omega$ represents the velocity of B, when A is stationary. Thus, the relative velocity of B with respect to A is $V_{BA} = R\omega$ and the total velocity is the sum of velocity of A and relative velocity of B with respect to A. This gives the relative velocity equation.

$$V_B = V_A + V_{BA} \tag{3.9}$$

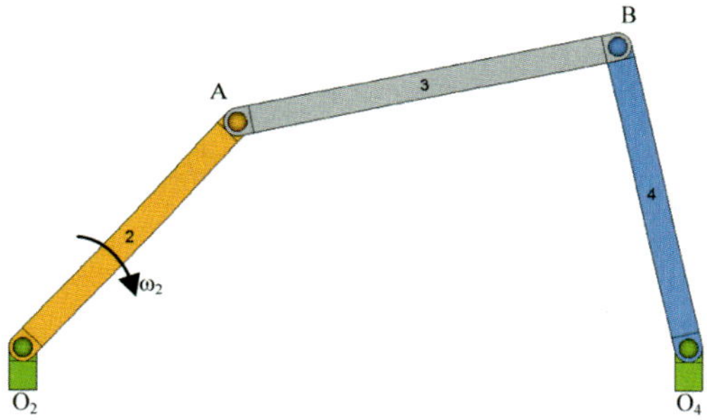

Fig. 3.12 A four-bar mechanism

As an example of velocity analysis, a four-bar mechanism is shown in Figure 3.12.

1. The velocity of point A is $V_A = \omega_2 \times O_2A$ as shown in Figure 3.13a.
2. Link 3 is shown separately in Figure 3.13b. Velocity of A is known in this link. Information about B is not known.
3. Link 4 is shown in Figure 3.13c which indicates that the velocity of B is in a direction perpendicular to O_4B, however its magnitude and sense are not known.
4. Apply the relative velocity equation to link 3. First draw the velocity O_va of point A to a suitable scale, see Figure 3.13d. Draw a line representing velocity of point B, perpendicular to link 4 from O_v.
5. Next draw a line representing the velocity of B with respect to A from a. The direction of this line is perpendicular to link 3. The direction of this relative velocity is shown in Figure 3.13e.
6. The two lines thus drawn from O_v and a meet at V in Figure 3.13d. This gives the velocity of point B in link 4.
7. The angular velocity of link 3 is $\omega_3 = V_{BA}/AB$ ccw (counter-clockwise).
8. The angular velocity of link 4 is $\omega_4 = V_B/O_4B$ cw (clockwise).

The solution for the following data is given in Figure 3.13f: $O_2O_4 = 8$ cm, $O_2A = 4$ cm, $AB = 4$ cm, $O_4B = 3$ cm, angle $AO_2O_4 = 60°$, $\omega_2 = 20$ rad/s cw.

Instead of just one position of the crank, we can perform the analysis for several positions; doing this manually takes a long time and a code like MotionSolve can achieve this in a short span of time, just a few minutes. This solution is also shown in Figure 3.13f.

Consider the slider crank mechanism in Figure 3.5. Let $O_2A = 4$ cm, $AB = 7$ cm and angle $AO_2O_4 = 45°$. The solution is given in Figure 3.5a.

The solution from MotionSolve is also given in Figure 3.5a.

Take another example and make a complete velocity analysis of a six-link mechanism in Figure 3.14.

1. The velocity of point A in link 2 is shown in Figure 3.15a.
2. Link 4 is shown in Figure 3.15b which indicates that the velocity of B is in the horizontal direction.

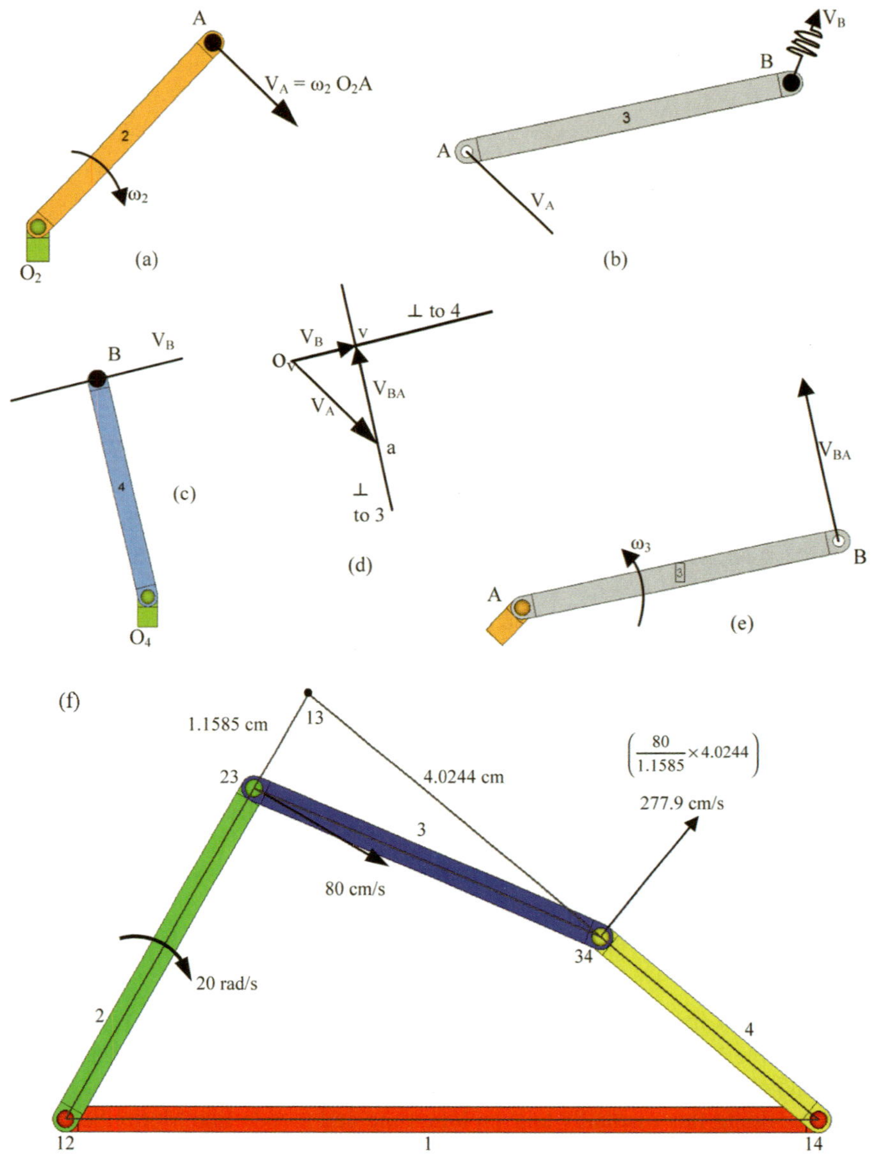

Fig. 3.13 (a)–(e) Velocity analysis of four-bar mechanism. (f) Example of four-bar mechanism

3. Link 3 is shown separately in Figure 3.15c. The velocity of *A* is known in this link.
4. Apply the relative velocity equation to link 3. First draw the velocity of point *A* to a suitable scale, see Figure 3.13d. Draw a horizontal line representing the velocity of point *B*.

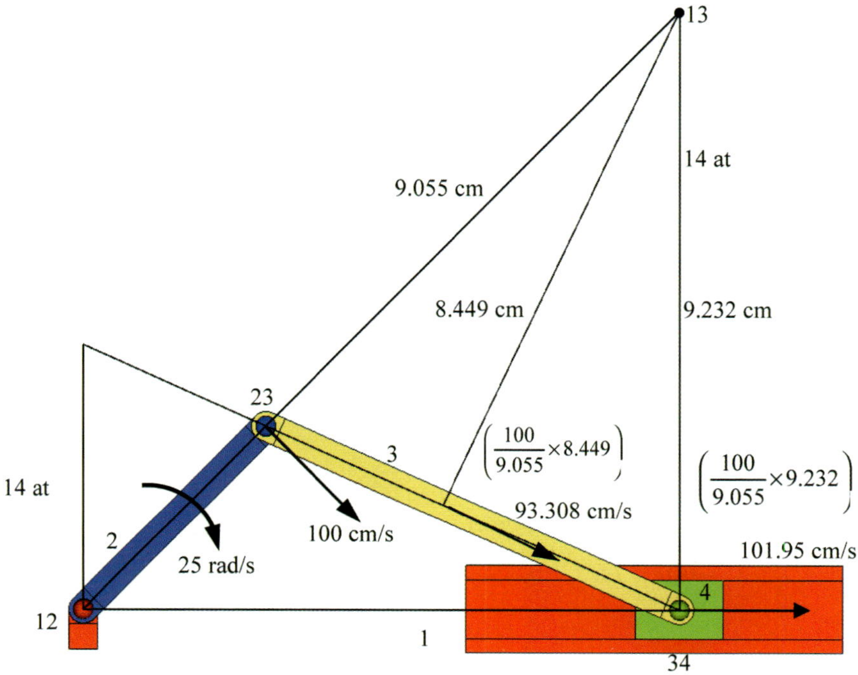

Fig. 3.5a Velocity solution of slider crank mechanism in Figure 3.5

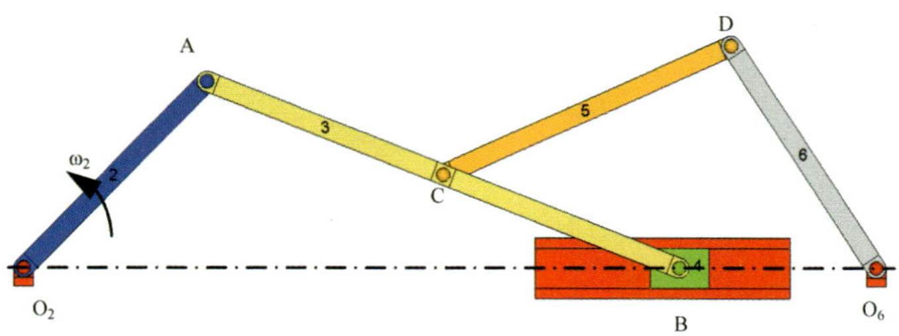

Fig. 3.14 A six-link mechanism

5. Next draw a line representing the velocity of B with respect to A. The direction of this line is perpendicular to link 3 as shown in Figure 3.15f.
6. The two lines thus drawn meet at a, see Figure 3.15g. This gives the velocity of link 4.
7. The angular velocity of link 3 is $\omega_3 = V_{BA}/AB$ cw.
8. Find the velocity of C by applying relative velocity equation $V_C = V_A + V_{CA}$, see Figures 3.15f and 3.15e.
9. Now, apply the relative velocity equation for link 5, $V_D = V_C + V_{DC}$.
10. The complete velocity diagram is given in Figure 3.15g.

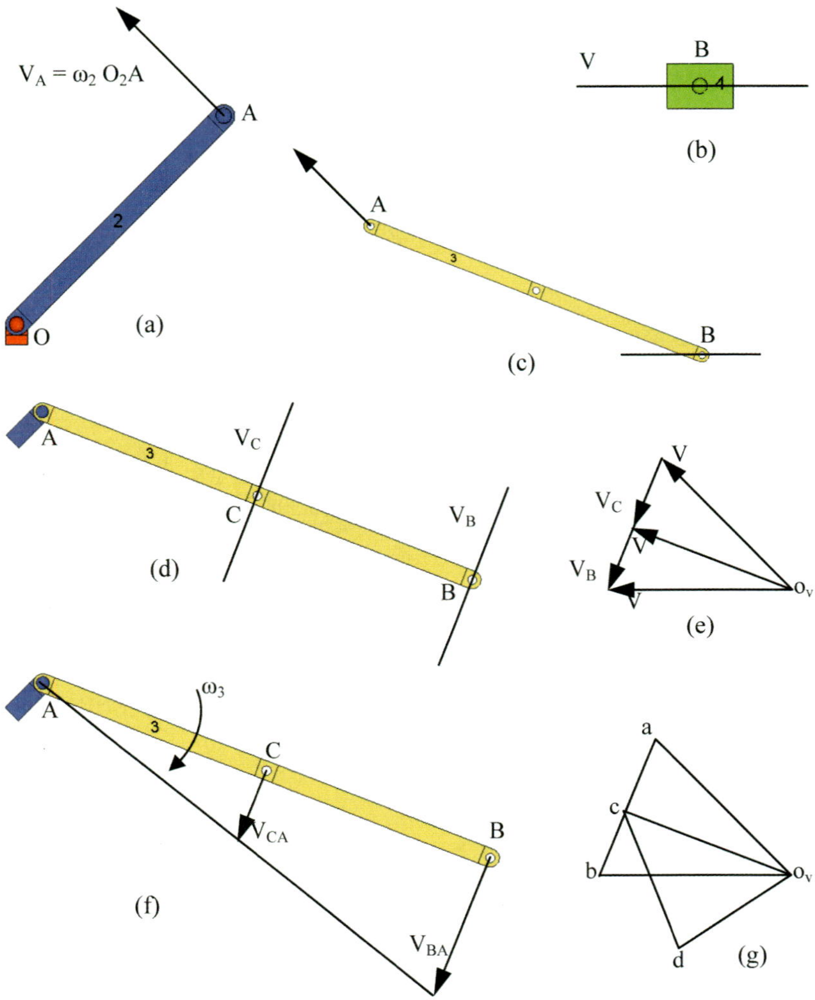

Fig. 3.15 Solution of six-link mechanism in Figure 3.14

3.2.3 Relative Velocity Equation of Two Coincident Points Belonging to Two Rigid Bodies

Consider two bodies M and N in plane OXY. Point A is the pole of body M having general plane motion in the OXY axis system given in Figure 3.16. Body N has relative motion with respect to body M (e.g., the slider block and the rocker arm in a quick return mechanism). P is a coincident point at the instant shown in Figure 3.16; specifically it is P_M in body M and P_N in body N. Since N has relative motion with respect to body M which we want to study, let us fix another axis system $A\xi\eta$ to the body M and the relative motion body N be determined in this moving frame with

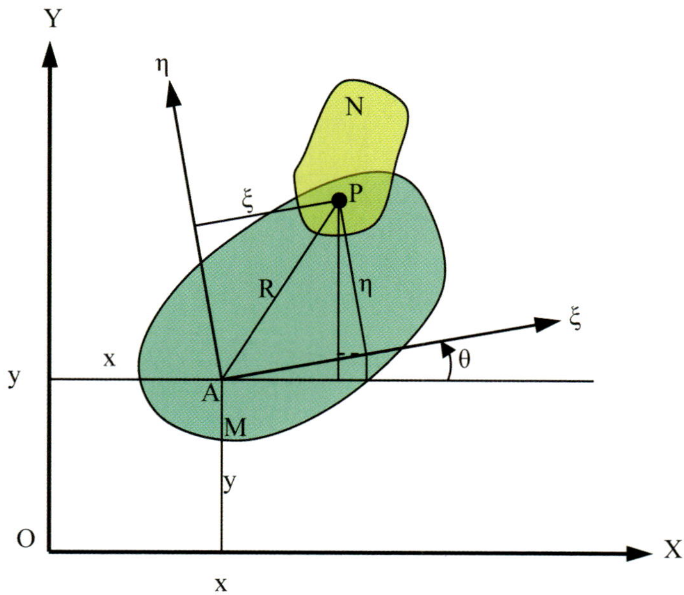

Fig. 3.16 Two coincident points, P, belonging to two rigid bodies M and N

respect to OXY. At the instant shown in Figure 3.16, $A\xi$ makes an angle θ with the positive direction of the X axis. The coordinates of point P_N are

$$X_P^N = X_A + \xi\cos\theta - \eta\sin\theta$$
$$Y_P^N = Y_A + \xi\sin\theta + \eta\cos\theta \qquad (3.10)$$

The corresponding velocities are

$$V_P^{NX} = V_A^X - \xi\omega\sin\theta + \frac{d\xi}{dt}\cos\theta - \eta\omega\cos\theta - \frac{d\eta}{dt}\sin\theta$$
$$V_P^{NY} = V_A^Y + \xi\omega\cos\theta + \frac{d\xi}{dt}\sin\theta - \eta\omega\sin\theta - \frac{d\eta}{dt}\cos\theta \qquad (3.11)$$

In the above $\omega = d\theta/dt$ is the angular velocity of body M, $U_\xi = d\xi/dt$ is the relative velocity of point P_N with respect to body M in the direction of ξ axis and $U_\eta = d\eta/dt$ is the relative velocity of point P_N with respect to body M in the direction of η axis. Rewriting

$$V_P^{NX} = V_A^X - \omega(\xi\sin\theta + \eta\cos\theta) + U_\xi\cos\theta - U_\eta\sin\theta$$
$$V_P^{NY} = V_A^Y + \omega(\xi\cos\theta - \eta\sin\theta) + U_\xi\sin\theta - U_\eta\cos\theta \qquad (3.12)$$

The total velocity is

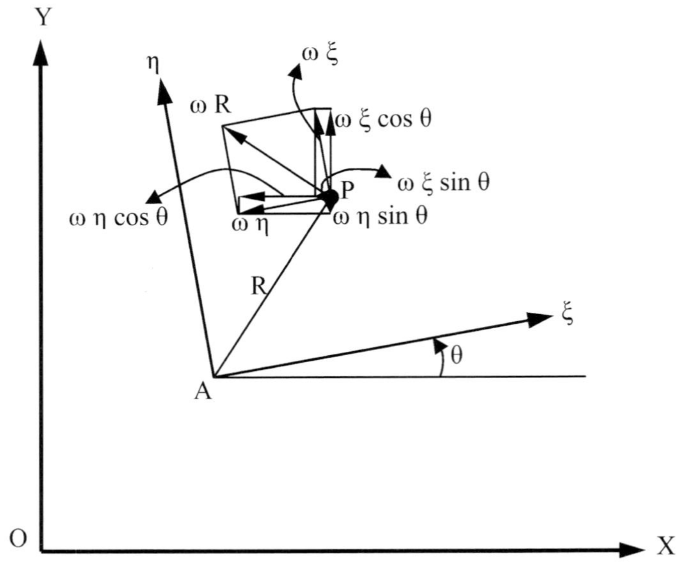

Fig. 3.17 Addition of vectors $[\omega(\xi\sin\theta + \rightarrow \xi\cos\theta) + \rightarrow \omega(\eta\cos\theta + \rightarrow \eta\sin\theta)]$

$$V_P^N = \left[V_A^X + \rightarrow V_A^Y\right] + \rightarrow [\omega\,(\xi\sin\theta + \rightarrow \xi\cos\theta) + \rightarrow \omega\,(\eta\cos\theta + \rightarrow \eta\sin\theta)]$$
$$+ \rightarrow \left[\left(U_\xi\cos\theta + \rightarrow U_\xi\sin\theta\right) + \rightarrow \left(U_\eta\sin\theta + \rightarrow U_\eta\cos\theta\right)\right] \tag{3.13}$$

The three rectangular bracketed quantities are identified as given below:

1. $[V_A^X + \rightarrow V_A^Y] = V_A$.
2. The addition of vectors in $[\omega(\xi\sin\theta + \rightarrow \xi\cos\theta) + \rightarrow \omega(\eta\cos\theta + \rightarrow \eta\sin\theta)]$ is shown in Figure 3.17.
3. The first term in this bracket gives $\omega(\xi\sin\theta + \rightarrow \xi\cos\theta) = \omega\xi$, which is parallel to the η axis and the second term is $\omega(\eta\cos\theta + \rightarrow \eta\sin\theta) = \omega\eta$ parallel to the ξ axis. The sum of these two terms gives $\omega\xi + \rightarrow \omega\eta = \omega R$ perpendicular to AP.
4. The addition of vectors in $[(U_\xi\cos\theta + \rightarrow U_\xi\sin\theta) + \rightarrow (U_\eta\sin\theta + \rightarrow U_\eta\cos\theta)]$ is shown in Figure 3.18. The result is $U_\xi + \rightarrow U_\eta = U$ making an angle $\varphi = \arctan U_\eta/U_\xi$ with the positive direction of ξ axis. This vector U is the total velocity of P_N relative to the body M and therefore it may be written as $U_{P_N P_M}$.

Therefore, equation (3.13) can be written as

$$V_P^N = V_A + \rightarrow \omega R + \rightarrow U_{P_N P_M} \tag{3.14}$$

Since $V_P^M = V_A + \rightarrow \omega R$, we can further write equation (3.14) as

$$V_P^N = V_P^M + \rightarrow U_{P_N P_M} \tag{3.15}$$

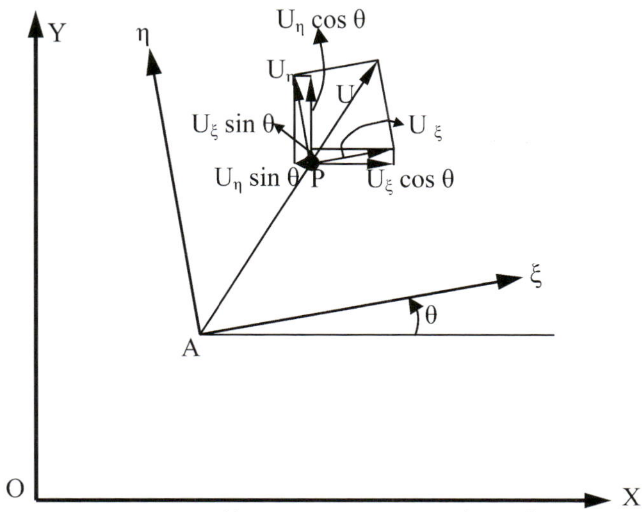

Fig. 3.18 Addition of vectors in $[(U_\xi \cos\theta + \rightarrow U_\xi \sin\theta) + \rightarrow (U_\eta \sin\theta + \rightarrow U_\eta \cos\theta)]$

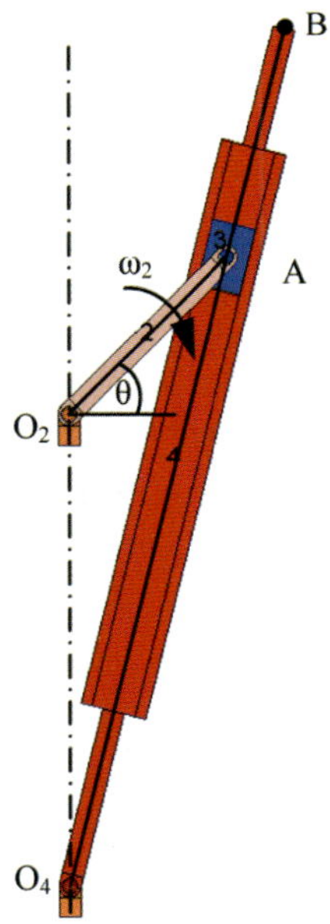

Fig. 3.19 Quick return mechanism

Now, let us make a complete velocity analysis of the quick return mechanism in Figure 3.19.

1. Figure 3.20a shows crank 2 with the slider 3 attached at point A. The velocity of A in link 2 is the same as that in link 3, $V_{A_3} = V_{A_2} = \omega_2 \times O_2A$.

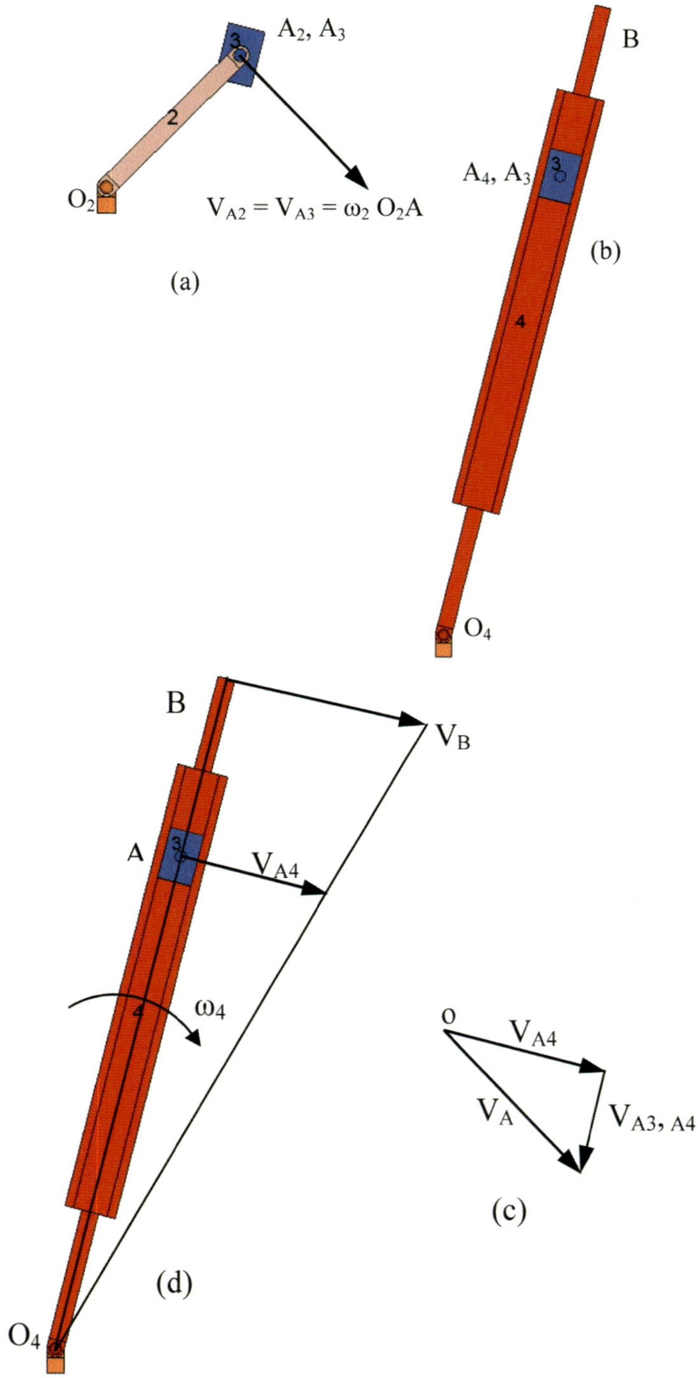

Fig. 3.20 Velocity analysis of quick return motion linkage

2. Link 4 is shown in Figure 3.20b along with link 3 for the configuration at this instant. We can identify link 3 to correspond to body N having relative motion with link 4 as body M. Therefore, equation (3.15) for this case is $V_{A_3} = V_{A_4} + \rightarrow V_{A_3A_4}$. Since link 4 rotates about O_4, the velocity of A_4 is perpendicular to the link and the relative motion of A_3 with respect to A_4 is along link 4.
3. Figure 3.20c gives this vector solution. (Note that the left-hand side of the equation is a known quantity.) The angular velocity of link 4 is $\omega_4 = V_{A_4}/O_4A_4$. The velocity of point B in link 4 can be scaled directly from the rotating link 4 shown in Figure 3.20d or determined from $V_B = \omega_4 \times O_4B$.

A velocity analysis is important for us in understanding how fast each link in a machine moves. We will learn later that high speed machines are advantageous in advanced applications like steam or gas turbines and space applications. But there is always a limitation on the speeds that we can use. In just the same way the displacement relations are used to determine velocities by taking a time derivative of the displacements; we determine accelerations by taking derivatives of velocities of the linkage. We will study later that the forces acting on machine members depend on the accelerations to which they are subjected (Newton's law). Larger accelerations mean larger forces, i.e., the stresses and the machine members should withstand these forces without any failure. Therefore the acceleration study becomes important.

The solution above is obtained for one position of the crank in its complete revolution of 360°. A designer would like to know the maximum values of velocity and acceleration and when they occur during a full revolution. The earlier practice was to repeat the above solution in, say, 5° intervals of the crank to make a plot of the kinematic quantities. This is time consuming and the modern computer codes allow us to determine the entire history in a very short turnover time, say a few minutes, once the model is ready.

Such a solution obtained in MotionSolve is shown in the avi of Figure 3.20 for the following data: $O_4O_2 = 45$ cm, $O_2C = 36$ cm, $O_2A = 14$ cm, $O_4B = 80$ cm, $BC = 25$ cm, and $\omega_2 = 200$ RPM.

3.3 Relative Acceleration Equation

The relative acceleration equation approach to determine accelerations is given here.

3.3.1 Rotation of a Rigid Link about a Fixed Axis

Consider the body M in Figure 3.21 rotating about a fixed axis O with angular velocity ω rad/s and angular acceleration $a = d\omega/dt$ rad/s^2, both measured positive in the ccw direction. The velocities of point B in the OXY axis system are given by, see equation (3.4)

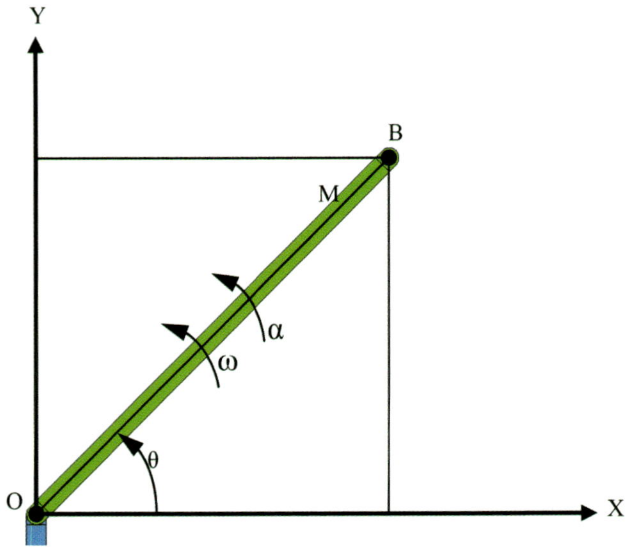

Fig. 3.21 Rotating body with angular acceleration

$$V_B^x = -R\omega \sin\theta \tag{3.16}$$

$$V_B^y = R\omega \cos\theta \tag{3.17}$$

Taking derivatives with respect to time, the accelerations are obtained.

$$\begin{aligned} a_B^X &= -R\omega^2 \cos\theta - R\alpha \sin\theta \\ a_B^Y &= -R\omega^2 \sin\theta + R\alpha \cos\theta \end{aligned} \tag{3.18}$$

The total acceleration of B is

$$a_B = a_B^X + \rightarrow a_B^Y = (R\omega^2 \cos\theta + \rightarrow R\omega^2 \sin\theta) + \rightarrow (R\alpha \sin\theta + \rightarrow R\alpha \cos\theta) \tag{3.19}$$

Addition of these vectors is shown in Figures 3.22 and 3.23. The first term in the above gives the radial (normal) component $\omega^2 R$ which is always directed towards the axis of rotation. The direction of tangential component αR from the second terms above is in the same sense of the angular acceleration. For α clockwise, the tangential component direction is illustrated in Figure 3.24.

3.3.2 Relative Acceleration of Two Points on a Rigid Body

Figure 3.25 shows the rigid body M. Velocities of point B were obtained earlier, see equation (3.7)

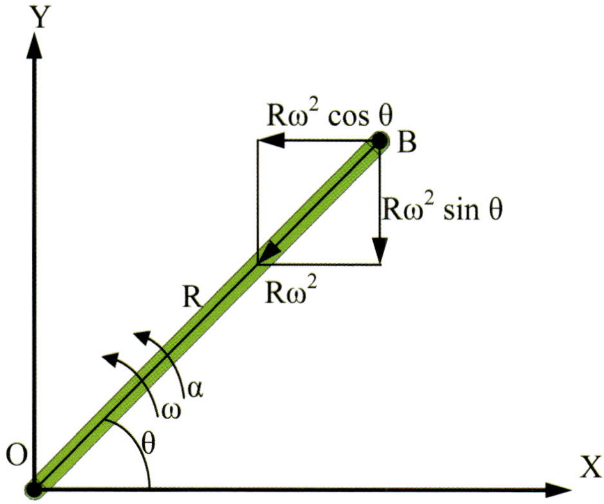

Fig. 3.22 Addition of $R\omega^2 \cos\theta + \rightarrow R\omega^2 \sin\theta$

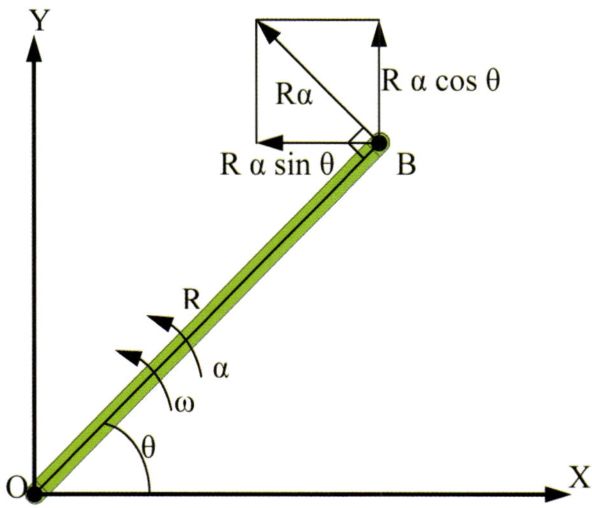

Fig. 3.23 Addition of $R\alpha \sin\theta + \rightarrow R\alpha \cos\theta$

$$
\begin{aligned}
V_B^x &= V_A^x - R\omega \sin\theta \\
V_B^y &= V_A^y + R\omega \cos\theta
\end{aligned}
\tag{3.20}
$$

The corresponding accelerations are

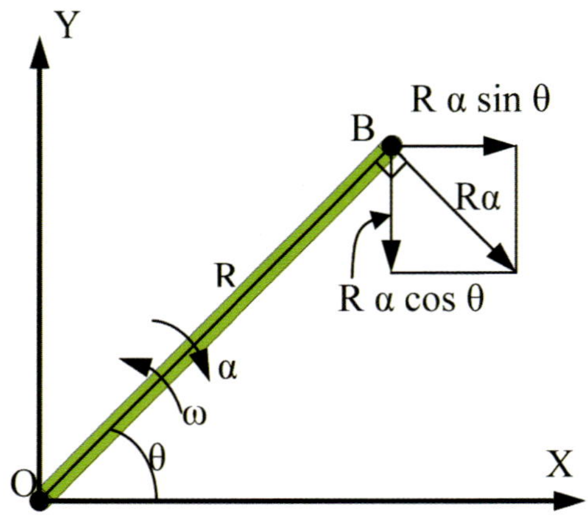

Fig. 3.24 Tangential component when acceleration is clockwise

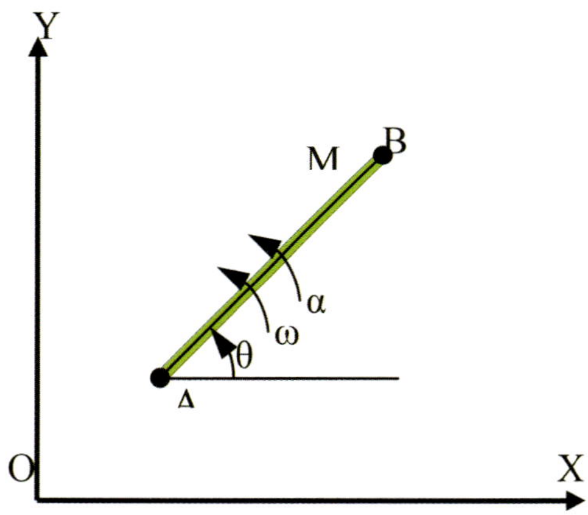

Fig. 3.25 Relative acceleration of two points on a rigid body

$$a_B^X = a_A^X - R\omega^2 \cos\theta - R\alpha \sin\theta$$
$$a_B^Y = a_A^Y - R\omega^2 \sin\theta + R\alpha \cos\theta \tag{3.21}$$

The total acceleration is

$$a_B = [a_A^X + \rightarrow a_A^Y] + \rightarrow [R\omega^2 \cos\theta + \rightarrow R\omega^2 \sin\theta]$$
$$+ \rightarrow [R\alpha \cos\theta + \rightarrow R\alpha \sin\theta] \tag{3.22}$$

Vectorial addition of the above three bracketed quantities is shown in Figure 3.26 and the relative acceleration equation is

$$a_B = a_A + a_{BA} \tag{3.23}$$

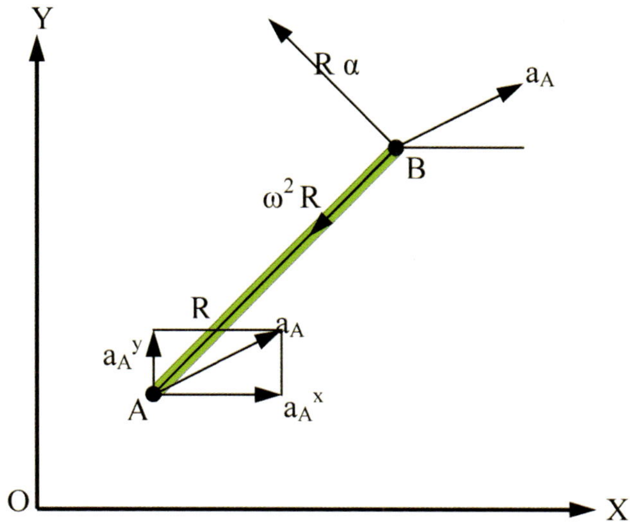

Fig. 3.26 Vectorial addition in equation (3.23)

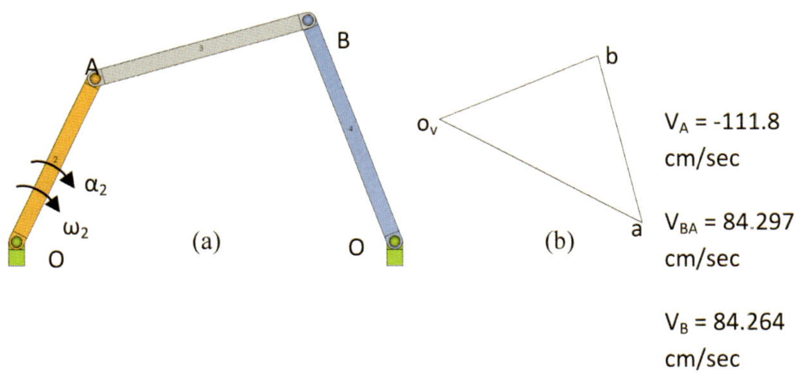

Fig. 3.27 Four-bar mechanism for acceleration analysis

Now make a complete acceleration analysis for the four-bar mechanism in Figure 3.27a. $O_2O_4 = 24.0$ cm, $O_2A = 11.18$ cm, $AB = 14.0$ cm, $O_4B = 14.68$ cm, $\angle AO_2O_4 = 63.435°$, $\omega_2 = 10$ rad/sec, $\alpha_2 = 10$ rad/sec^2.

1. The velocity diagram is obtained earlier and shown again in Figure 3.27b.
2. Figure 3.28a gives accelerations in the normal and tangential directions of link 2.
3. Link 3 is shown in Figure 3.28b. The acceleration of point A only is known at this stage.
4. Link 4 is shown in Figure 3.28c. The normal component is known, since its angular velocity is determined earlier. The tangential component is perpendicular to link 4; its direction however is not known at this stage.

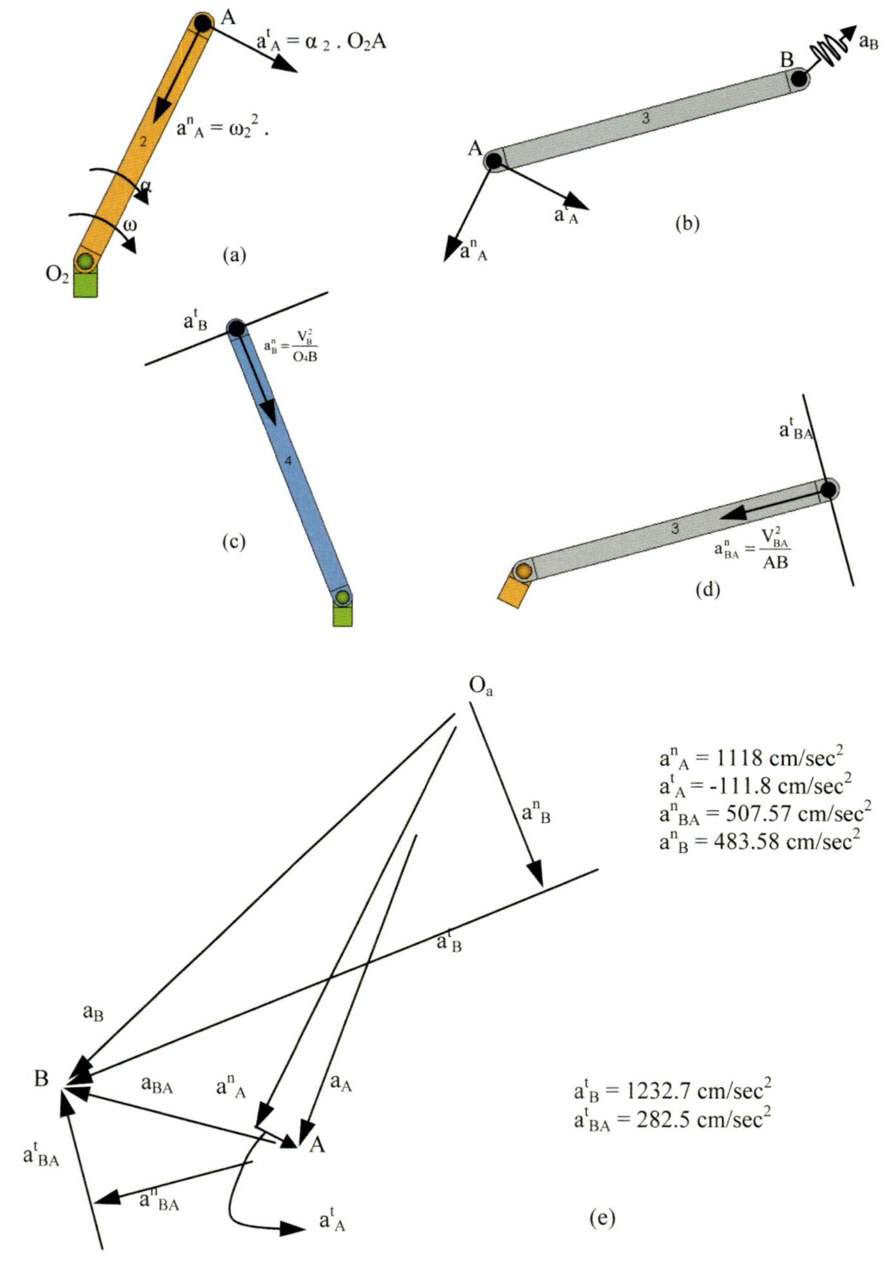

Fig. 3.28 Acceleration analysis of the mechanism in Figure 3.27

5. Link 3 is shown again in Figure 3.28d, where the relative normal component of acceleration is known. The relative tangential component of acceleration is not known at this stage.

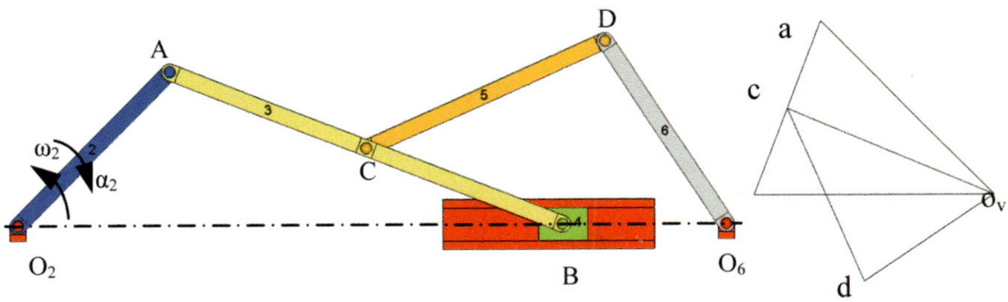

Fig. 3.29 Six-link mechanism with velocity diagram

6. The relative acceleration equation for link 3 is $a_B^n + a_B^t = a_A^n + A_A^t + a_{BA}^n + a_{BA}^t$. In this, the quantities a_B^n, a_A^n, a_A^t, a_{BA}^n are completely defined. The unknowns are. Their magnitudes are not known, however, their directions are known. Hence, the acceleration diagram in Figure 3.28e can be drawn.
7. First draw $a_A = a_A^n + a_A^t$ as shown to get the point A on the acceleration diagram.
8. Draw a_{BA}^n the normal component of acceleration of B with respect to A from point A on the acceleration diagram. At the end of this component, draw a line (perpendicular to link 3) representing the tangential component of acceleration of B with respect to A.
9. Now, from O_a draw a_B^n, the normal component of acceleration of B in link 4. At the end of this component, draw a line (perpendicular to link 4) representing the tangential component of acceleration of B in link 4. This line intersects the line drawn in step 8 above at B on the acceleration diagram.
10. From this acceleration diagram, the angular accelerations of links 3 and 4 can be obtained.

The solution from MotionSolve for the crank rotation from 63° to 3° is shown in Figure 3.28.

Next consider the six-link mechanism with the velocity diagram shown in Figure 3.29. Make a complete acceleration analysis.

1. Figure 3.30a gives the accelerations in the normal and tangential directions of link 2. Note that the angular acceleration of link 2 is clockwise.
2. Link 3 is shown in Figure 3.30b. Acceleration of point A only is known.
3. Link 4 is a slider; therefore its total velocity is in the horizontal direction as in Figure 3.30c. This is indicated in Figure 3.30b, its magnitude is yet to be determined.
4. Link 3 is shown again in Figure 3.30d, where the relative normal component of acceleration is known. The relative tangential component of acceleration is not known at this stage.
5. The relative acceleration equation for link 3 is $a_B = a_A^n + a_A^t + a_{BA}^n + a_{BA}^t$. In this, the only unknowns are a_B, a_{BA}^t. Their magnitudes are not known, however, their directions are known. Hence, the acceleration diagram in Figure 3.30e can be drawn.

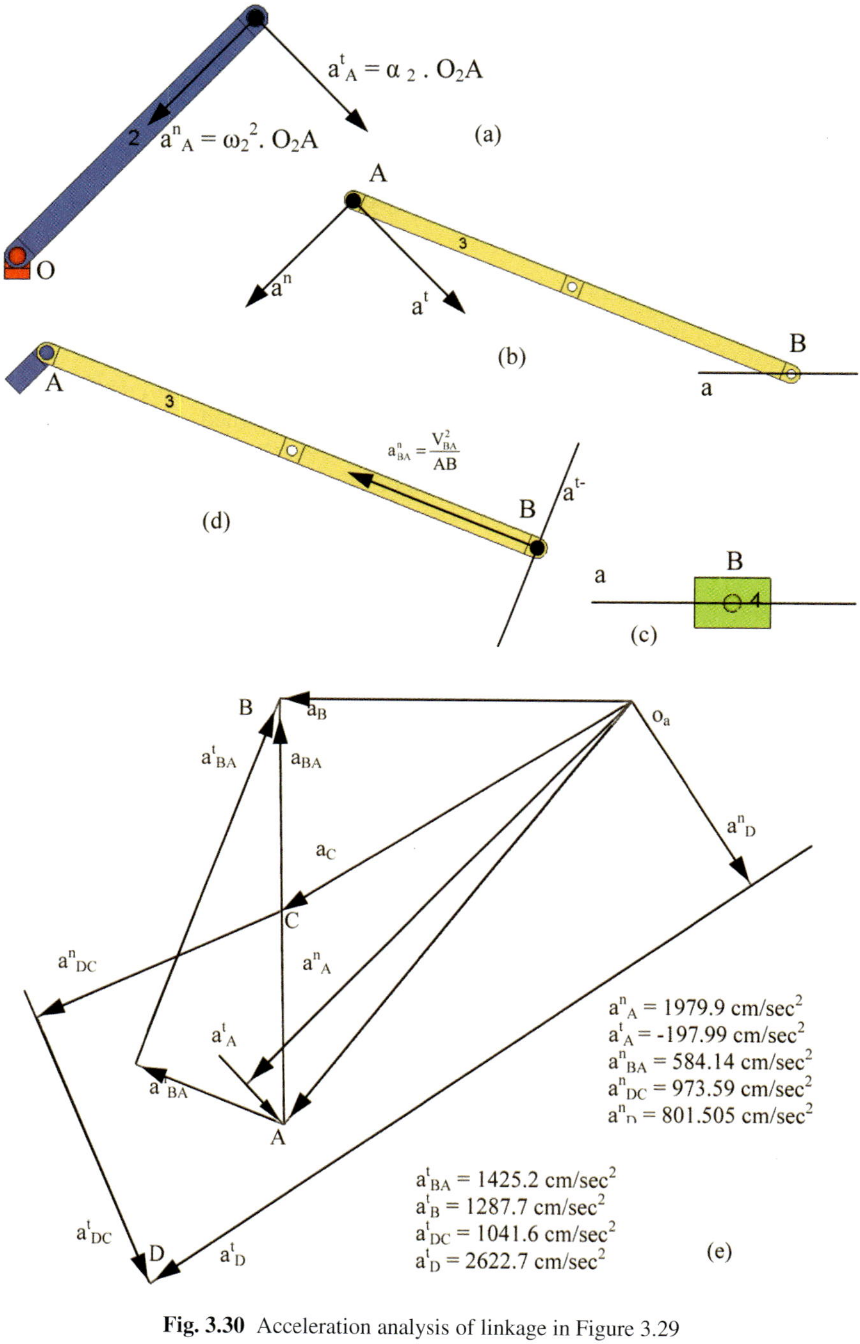

Fig. 3.30 Acceleration analysis of linkage in Figure 3.29

6. Draw $a_A = a_A^n + a_A^t$ as shown to get the point A on the acceleration diagram.
7. Draw the normal component of acceleration of B with respect to $A$from point A on the acceleration diagram. At the end of this component, draw a line (perpendicular to link 3) representing the tangential component of acceleration of B with respect to A.
8. Now, from O_a draw a horizontal line representing the acceleration of B in link 4. This line intersects the line drawn in step 7 at B on the acceleration diagram.
9. From this acceleration diagram, the angular accelerations of links 3 and 4 can be obtained.
10. Acceleration of B with respect to A is given by AB on the acceleration diagram. Acceleration of C with respect to A is obtained by the ratio of AC to AB of link 3 and thus locates point C on the acceleration diagram.
11. Now, relative acceleration equation $a_D^n + a_D^t = a_C + a_{DC}^n + a_{DC}^t$ for link 5 is applied. Here, a_D^n, $a_C + a_{DC}^n$ are completely defined. The unknowns are the magnitudes of a_D^t, a_{DC}^t, their lines of action are known. Therefore, the acceleration diagram can be completed as shown in Figure 3.30e.

The solution from MotionSolve for the crank rotation from 45° to 70° is shown in Figure 3.30.

3.3.3 Relative Acceleration Equation of Two Coincident Points Belonging to Two Rigid Bodies

The relative velocity equation of two coincident points P_M and P_N considered earlier is extended to acceleration analysis. Figure 3.31 shows the two bodies, body N having relative motion with respect to body M. Differentiating equations (3.12) we get

$$\begin{aligned} a_P^{NX} &= a_A^X - \omega^2(\xi \cos\theta - \eta \sin\theta) - a(\xi \sin\theta + \eta \cos\theta) \\ &\quad + (a_\xi \cos\theta - a_\eta \sin\theta) - 2U_\xi \omega \sin\theta - 2U_\eta \omega \cos\theta \\ a_P^{NY} &= a_A^Y - \omega^2(\xi \sin\theta + \eta \cos\theta) + a(\xi \cos\theta - \eta \sin\theta) \\ &\quad + (a_\xi \sin\theta + a_\eta \cos\theta) + 2U_\xi \omega \cos\theta - 2U_\eta \omega \sin\theta \end{aligned} \tag{3.24}$$

In the above $a_\xi = dU_\xi/dt$, $a_\eta = dU_\eta/dt$ are the relative accelerations of point P_N with respect to the body M in ξ and η directions respectively. Vector addition of all the components in equation (3.24) gives the total acceleration of P_N.

1. $a_A^X + \!\rightarrow a_A^Y$ is the total acceleration of the pole A in body M.
2. $\omega^2(\xi \cos\theta + \!\rightarrow \eta \sin\theta) + \!\rightarrow \omega^2(\xi \sin\theta + \!\rightarrow \eta \cos\theta)$: Addition of these vectors is shown in Figure 3.32. This is identified as the normal component of acceleration of $P_M = \omega^2 R$.

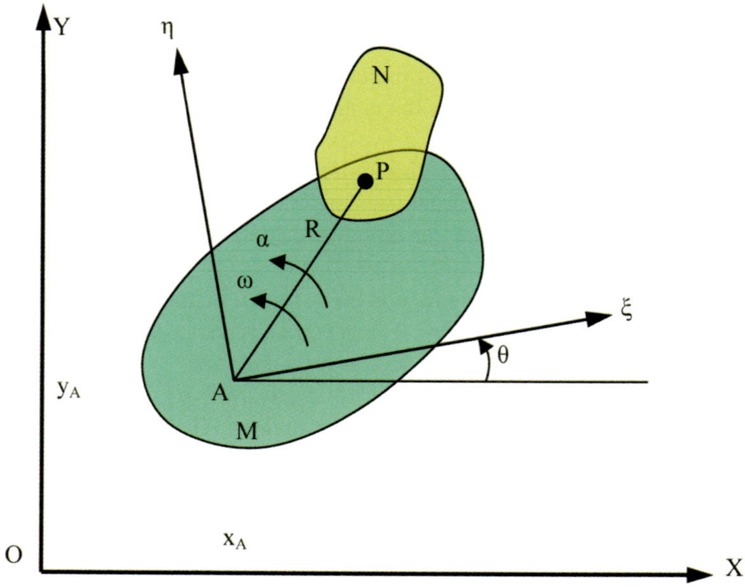

Fig. 3.31 Acceleration equation of two coincident points in two rigid bodies

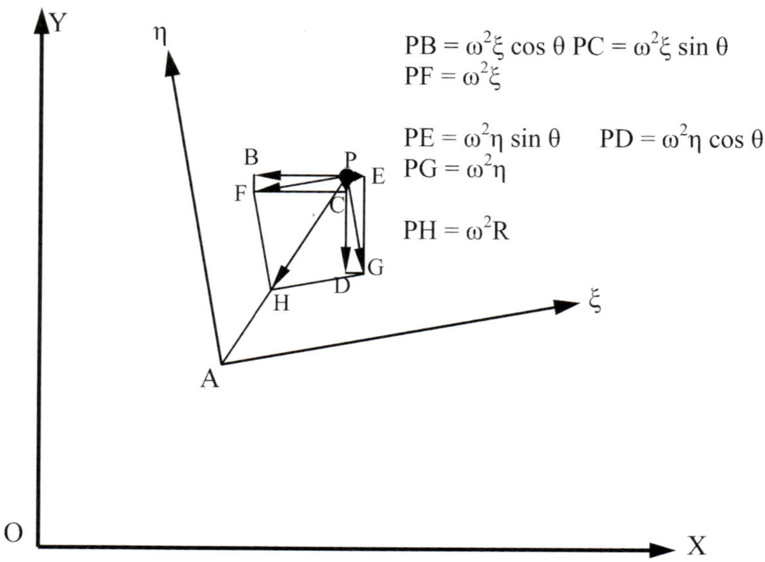

Fig. 3.32 Addition of vectors $\omega^2(\xi \cos\theta + \rightarrow \eta \sin\theta) + \rightarrow \omega^2(\xi \sin\theta + \rightarrow \eta \cos\theta)$

3. $a(\xi \sin\theta + \rightarrow \xi \cos\theta) + \rightarrow a(\eta \sin\theta + \rightarrow \eta \cos\theta)$: Figure 3.33 gives the addition of these vectors giving the tangential component of acceleration of $P_M = R\alpha$.

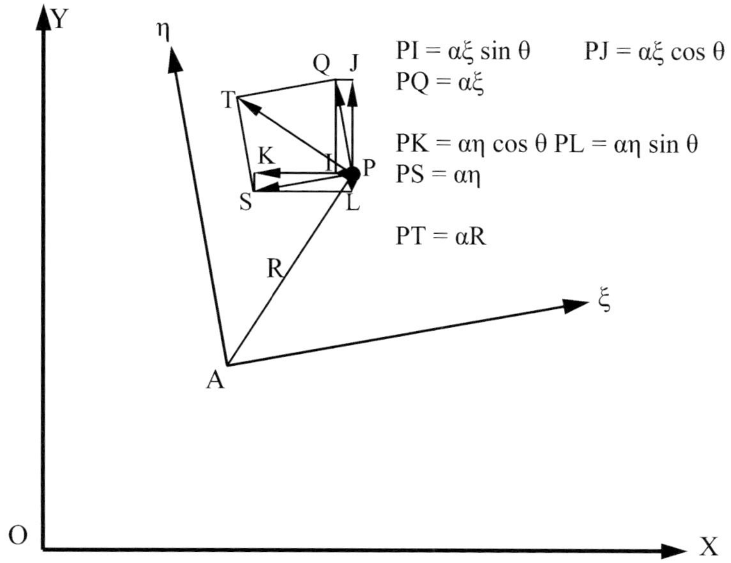

Fig. 3.33 Addition of vectors $a(\xi \sin\theta + \rightarrow \xi \cos\theta) + \rightarrow a(\eta \sin\theta + \rightarrow \eta \cos\theta)$

4. $(a_\xi \cos\theta + \rightarrow a_\xi \sin\theta) + \rightarrow (a_\eta \cos\theta + \rightarrow a_\eta \sin\theta)$: Figure 3.34 gives the addition of these terms giving the relative acceleration a_r of P in the body N with P in the body M which is equal to $(a_\xi + a_\eta)$.
5. $2U_\xi\omega \cos\theta + \rightarrow 2U_\xi\omega \sin\theta$: The addition of these two terms gives $2U_\xi\omega$ parallel to the η axis as shown in Figure 3.35.
6. Addition of $2U_\eta\omega \cos\theta + \rightarrow 2U_\eta\omega \sin\theta$ gives $2U_\eta\omega$ parallel to the ξ axis. The sum of the two vectors $2U_\xi\omega$ and $2U_\eta\omega$ gives a vector of magnitude $2U\omega$ making an angle $\beta = \arctan U_\eta/U_\xi$ with the positive direction of the η axis. Since the relative velocity vector makes an angle $\phi = \arctan U_\eta/U_\xi$ with the positive direction of the ξ axis, the $2U\omega$ vector is perpendicular to the relative velocity vector.

The total acceleration of all components can now be summarized as

$$a_P^N = a_A + \rightarrow \omega^2 R + \rightarrow \alpha R + \rightarrow a_r + 2U\omega \tag{3.25}$$

This can be further simplified as

$$a_P^N = a_P^M + \rightarrow a_{P_N P_M} + 2U\omega \tag{3.26}$$

The relative acceleration equation of two coincident points thus shows an additional term $2U\omega$, if we compute the acceleration of P in the body N to consist of acceleration of P in the body M plus the relative acceleration of P in the body N with respect to P in the body M. The component $PC = 2U\omega$ is the Coriolis component of acceleration. PC' is for the clockwise ω direction. The direction of

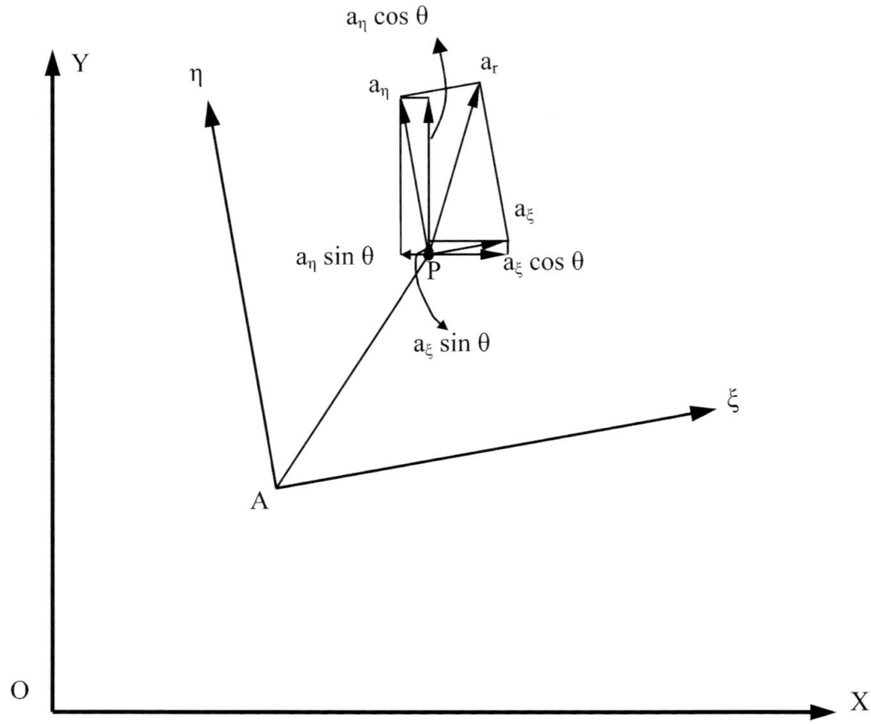

Fig. 3.34 Addition of vectors $(a_\xi \cos\theta + \rightarrow a_\xi \sin\theta) + \rightarrow (a_\eta \cos\theta + \rightarrow a_\eta \sin\theta)$

this vector can be determined by a simple rule as illustrated in Figure 3.36. At the base of relative velocity vector U, draw a circle going in the direction of the angular velocity ω and proceed in a direction perpendicular to U when it meets with the relative velocity vector.

The quick return mechanism considered earlier is given in Figure 3.37 along with the velocity diagram. Make a complete acceleration analysis.

1. Acceleration of A in link 2 or 3 is given by $a^n_A = \omega_2^2 \times O_2A$ and $a^t_A = \alpha_2 \times O_2A$ as shown in Figure 3.38a.
2. Link 4 with slider 3 is shown in Figure 3.38b. Recall that link 4 is body M and link 3 is body N.
3. Therefore, $a^n_{A_3} + \rightarrow a^t_{A_3} = a^n_{A_4} + \rightarrow a^t_{A_4} + \rightarrow a_{A_3A_4} + \rightarrow 2V_{A_3A_4}\omega_4$. In this, $a^n_{A_3}$, $a^t_{A_3}$ are completely known. $a^n_{A_4} = \omega_4^2 \times O_4A$ is also a known vector. The Coriolis component of acceleration $2V_{A_3A_4}\omega_4$ can also be determined completely with the help of Figure 3.38c. The only unknown quantities are $a^t_{A_4}$, $a_{A_3A_4}$. Their directions are however known and therefore, the acceleration diagram can be drawn as shown in Figure 3.38d.
4. First draw the sum of normal and tangential components of acceleration of A_2 or A_3 to give O_aA in Figure 3.38d.

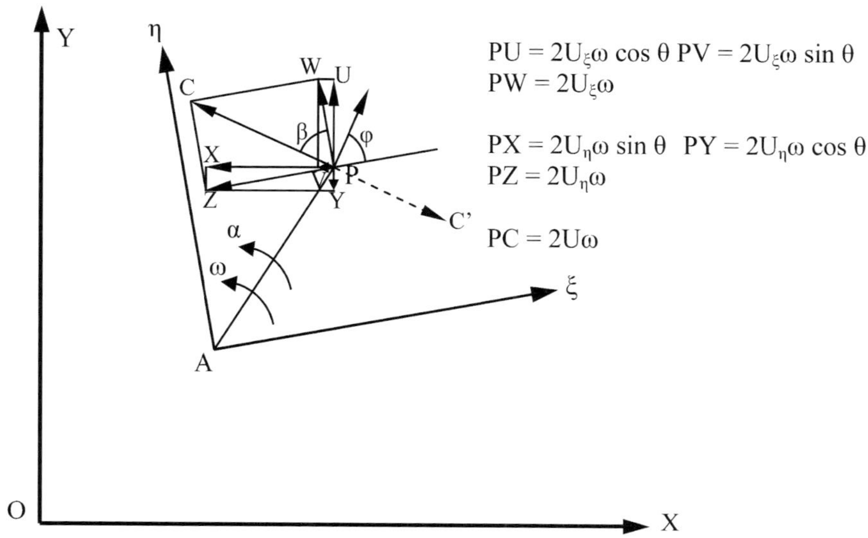

Fig. 3.35 Addition of vectors $2U_\xi a\cos\theta + \rightarrow 2U_\xi\omega\sin\theta$ and $2U_\eta\omega\cos\theta + \rightarrow 2U_\eta\omega\sin\theta$

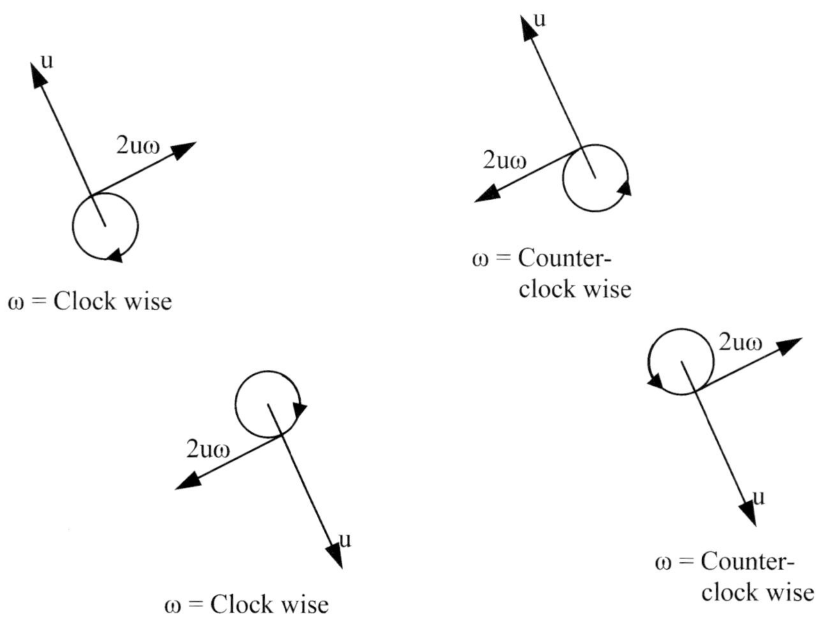

Fig. 3.36 Coriolis component direction

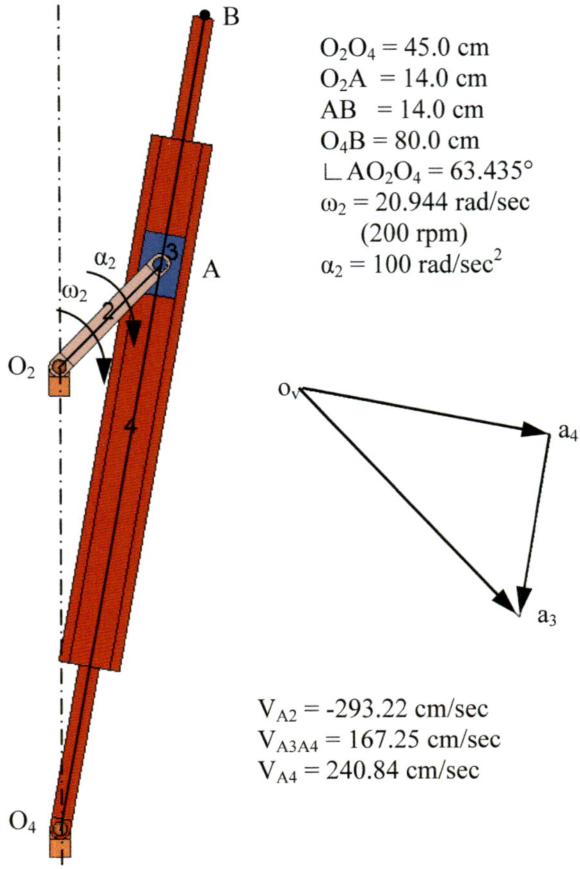

Fig. 3.37 Quick return motion mechanism with velocity diagram

5. Next draw the normal component of acceleration of A_4 from O_a. Note that this term is on the right-hand side of the relative acceleration equation. A line perpendicular to this component is drawn to represent the tangential component of acceleration of A_4. The magnitude of this line is not known at this stage.
6. At the tip of acceleration vector of A_3 set up a vector corresponding to the Coriolis acceleration $2V_{A_3A_4}\omega_4$ in the direction given by Figure 3.38c. Note that this component is on the right-hand side of the relative acceleration equation and therefore this vector should end at point A_3 to give the total acceleration of A_3.
7. At the foot of the Coriolis vector, draw a line corresponding to the relative acceleration $a_{A_3A_4}$, which is along the link 4. This intersects the tangential component of acceleration of A_4 and the acceleration diagram is now completed.
8. Produce the line $O_aa_{A_4}$ in the same ratio as that of Figure 3.38b to give the total acceleration of B.

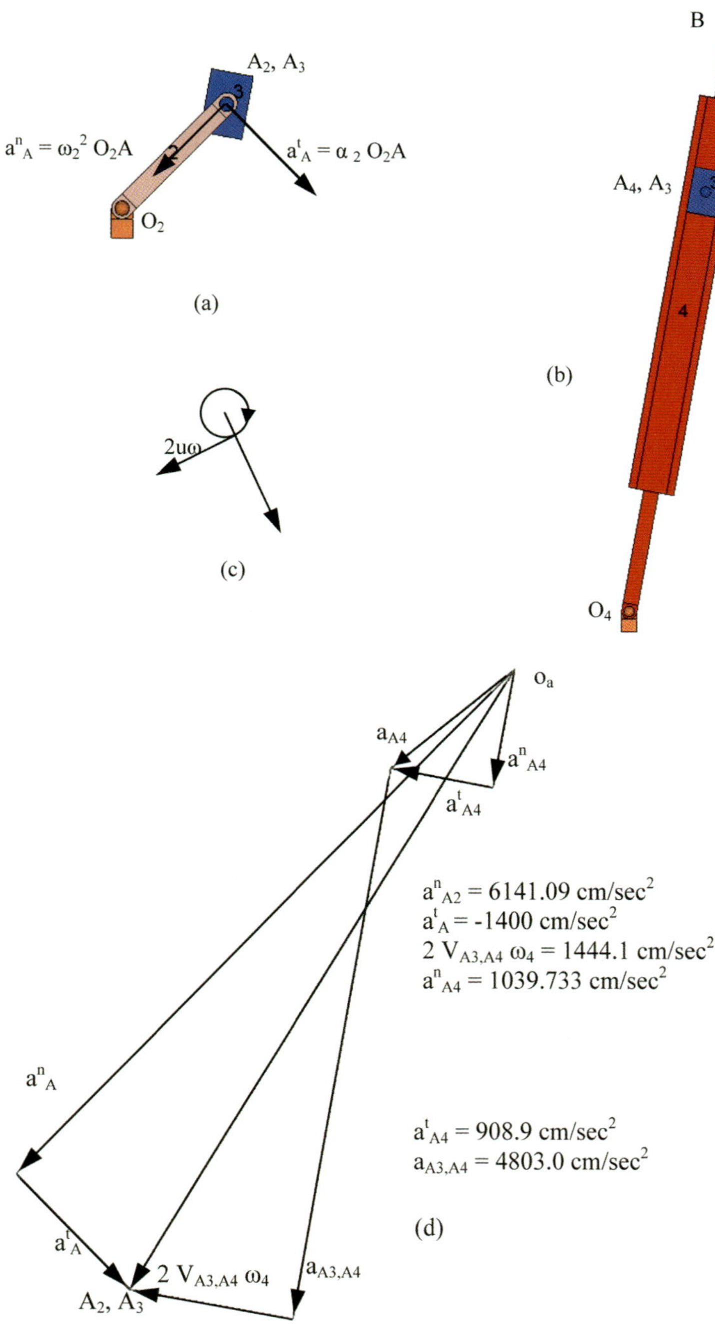

Fig. 3.38 Acceleration solution of quick return motion mechanism

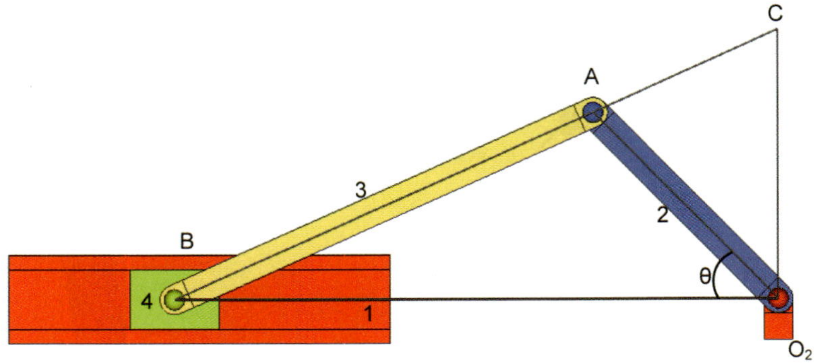

Fig. 3.39 Klein's construction – velocity analysis

3.4 Acceleration Analysis of Reciprocating Engine Mechanism

Historically, the reciprocating steam engine was the main source of power in the 19th century. The reciprocating internal combustion engine began replacing the steam engine in the 20th century. Thus, a reciprocating machine was the most common power source for nearly over two centuries. Therefore, designers had to develop special graphical methods prior to the present computer era which began only in the last three to four decades. We will study some of these methods to determine the velocity and acceleration of reciprocating engine mechanism members when the crank is rotating at constant speed.

3.4.1 Klein's Construction

Figure 3.39 shows the crank making an angle θ with the line of stroke. The velocity diagram of the mechanism is given by O_2AC where C is the point obtained by the intersection of line BA extended and the perpendicular to the line of stroke at O_2. O_2C represents the magnitude of piston velocity to the scale O_2A representing the crank velocity and AC represents the relative velocity of B with respect to A. This can be verified from the velocity diagram of the engine in Figure 3.40a and the triangle O_2AC in Figure 3.40b.

Figure 3.41 shows the construction of acceleration diagram O_2AQP. The following steps are used in this construction:

1. Draw a circle with AB as diameter.
2. Draw a circle with A as center and AC as radius.
3. The two circles intersect at L and M.
4. Line LM intersects the connecting rod at Q and the line of stroke at P.

Triangles ALQ and ALB are similar, therefore

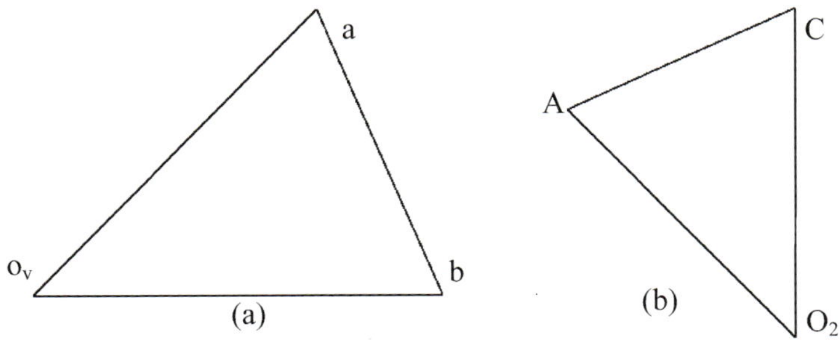

Fig. 3.40 Velocity diagram

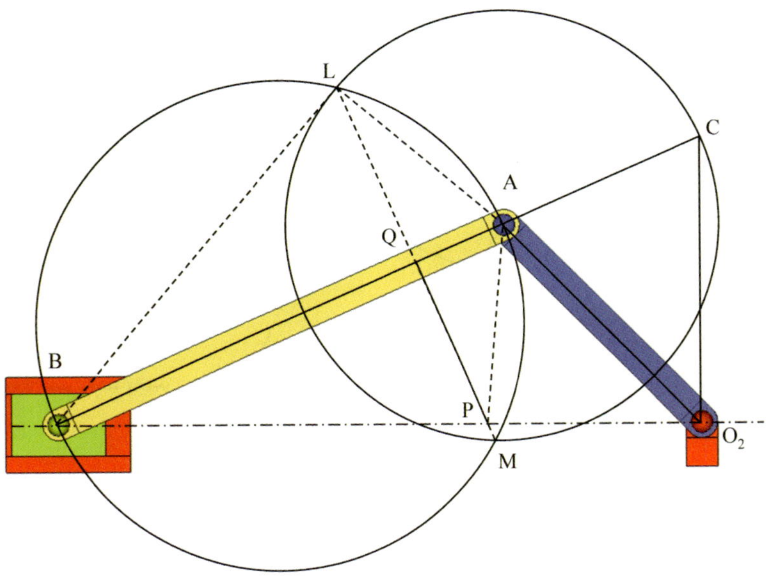

Fig. 3.41 Klein's construction for acceleration

$$\frac{AQ}{AL} = \frac{AL}{AB}$$

$$\therefore AQ = \frac{AL^2}{AB} \tag{3.27}$$

Since $AL = AC$

$$AQ = \frac{AC^2}{AB} \tag{3.28}$$

Therefore AQ corresponds to the normal component acceleration a^n_{BA} of B with respect to A to a scale O_2A equal to $\omega_2^2 \times O_2A$.

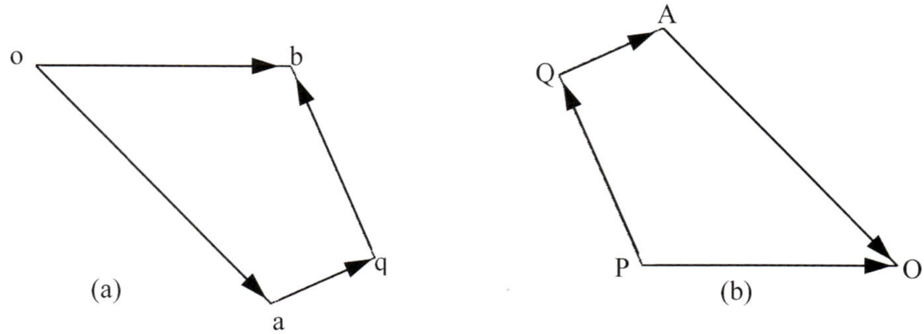

Fig. 3.42 Acceleration diagram

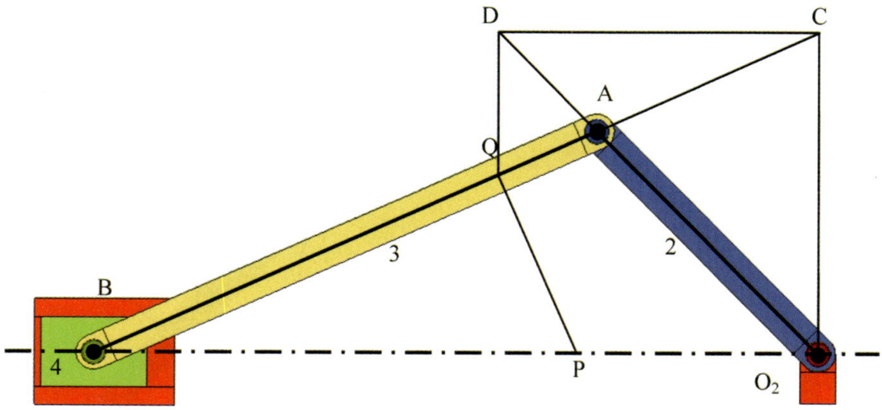

Fig. 3.43 Ritterhaus construction

Figure 3.42a shows the acceleration diagram drawn to the same scale as above by using the relative acceleration equation. This acceleration diagram can be redrawn as shown in Figure 3.42b (check the vectorial addition) and obtain an identical diagram O_2PQA as in Figure 3.41. O_2P gives the acceleration of the piston.

3.4.2 Ritterhaus Construction

Figure 3.43 illustrates this construction procedure.

1. Erect a perpendicular at O_2 to meet BA at C.
2. Draw a horizontal line from C to meet O_2A at D.
3. Drop a perpendicular from D to intersect AB at Q.
4. Draw a line from Q perpendicular to AB. This line intersects the line of stroke at P.

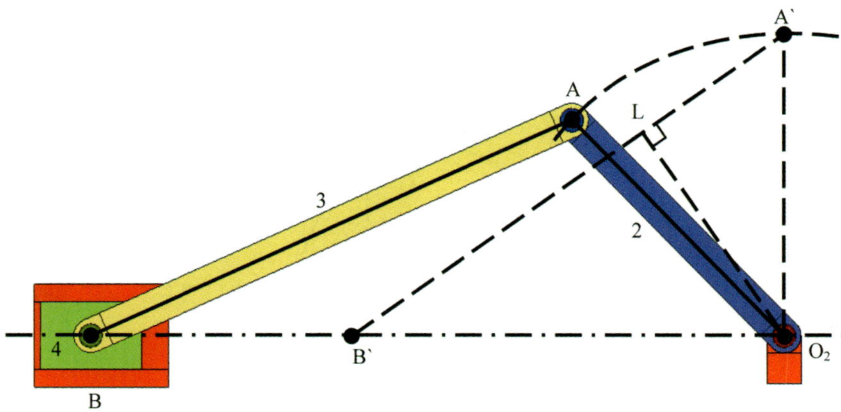

Fig. 3.44 Bennet's construction

From the two sets of similar triangles O_2AC & ADQ and ACD & O_2AB, we can find that equation (3.28) is satisfied. Therefore, O_2PQA in Figure 3.43 is the acceleration diagram.

3.4.3 Bennet's Construction

The following steps describe the procedure:

1. Draw the mechanism configuration when the crank is at 90°, $O_2A'B'$ as in Figure 3.44.
2. Erect a perpendicular O_2L to the line $A'B'$ and measure the length $A'L$.
3. Locate this point L on the connecting rod in Figure 3.45.
4. Draw a perpendicular to the connecting rod from L to meet the line of stroke at M.
5. Erect a vertical line from M to meet the connecting rod at Q.
6. Draw a perpendicular to the connecting rod from Q to meet the line of stroke at P.
7. O_2PQA is the required acceleration diagram.

3.5 Analytical Determination of Velocity and Acceleration of the Piston

We notice that the piston goes through two dead centers for every revolution of the crank. Its velocity becomes zero when the piston reaches top dead center and bottom dead center. In these two positions, it has to reverse its motion and accelerate and

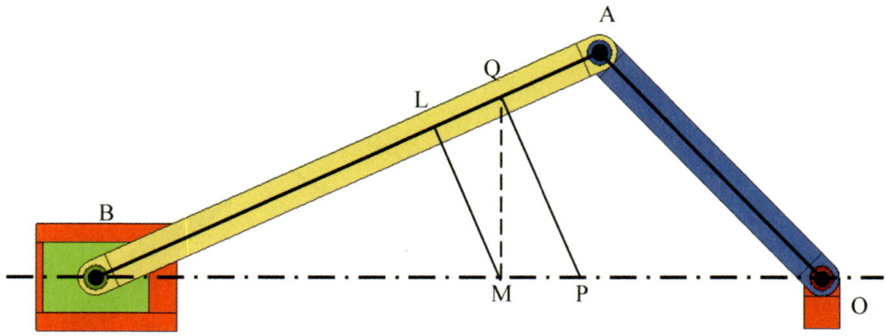

Fig. 3.45 Bennet's construction

then decelerate to reach the other dead center. This makes a reciprocating engine not ideally suited for a good design. But we lived with the reciprocating steam engine for more than a century before abandoning it in favor of the steam turbine that became the most extensively used from the beginning of the 20th century. For an internal combustion engine, complex crank arrangements and cylinders were devised leading to multicylinder engines that helped designers to raise speeds, get better efficiencies and also balance the piston forces in every revolution. Thus the study of piston acceleration became important in the study of Theory of Machines. The graphical methods widely used in the 19th century were replaced by faster and more accurate analytical methods.

The piston is at a distance x from the main bearing center, when the crank is making an angle θ and the connecting rod an angle φ with the line of stroke as shown in Figure 3.46. The radius of the crank is r and the length of the connecting rod is l. At the outer dead center position of the crank, the piston is in the position B' at distance $r + l$ from the main bearing center.

$$x = r\cos\theta + l\cos\varphi \tag{3.29}$$

Note that $r\sin\theta = l\sin\varphi$ and, with n representing the ratio of connecting rod to crank l/r

$$\sin\varphi = \frac{1}{n}\sin\theta$$

$$\cos\varphi = \sqrt{1 - \frac{\sin^2\theta}{n^2}} \tag{3.30}$$

Therefore

$$x = r\cos\theta + l\sqrt{1 - \frac{\sin^2\theta}{n^2}} \tag{3.31}$$

Differentiating, the velocity of the piston is

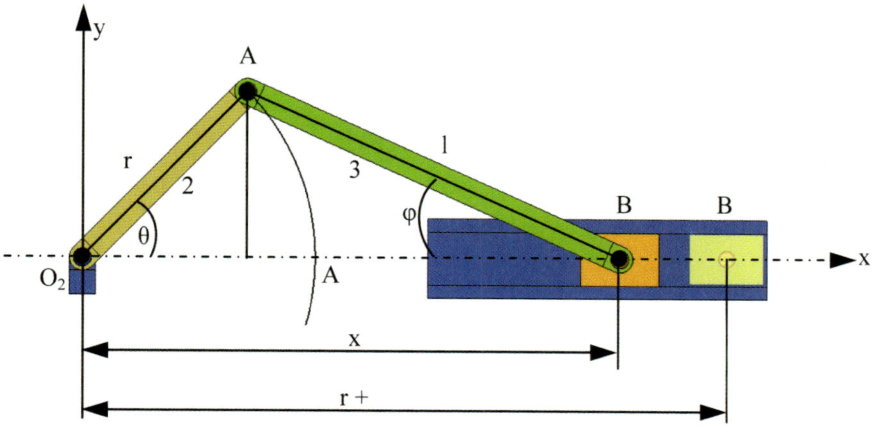

Fig. 3.46 Analytical derivation for acceleration

$$v = -r\omega\left[\sin\theta + \frac{1}{2n}\left(1 - \frac{\sin^2\theta}{n^2}\right)^{-\frac{1}{2}}\sin 2\theta\right] \tag{3.32}$$

Differentiating once more, the acceleration of the piston is

$$a = -r\omega^2\left[\cos\theta + \frac{(n^2-1)\cos 2\theta + \cos^4\theta}{n^3\left(1 - \frac{\sin^2\theta}{n^2}\right)^{\frac{3}{2}}}\right] \tag{3.33}$$

The above expression for piston acceleration is unwieldy for early day engineering use. Also the engineers would like to know the velocity and piston as harmonic components with the engine speed rather than exact periodically varying kinematic quantities. We therefore use a harmonic analysis and find the required velocity and acceleration components.

3.5.1 Harmonic Analysis for Velocity and Acceleration of the Piston

Equations (3.32) and (3.33) are exact expressions for velocity and acceleration of the piston. We can see that they are periodic functions of the crank angle θ. Using the Binomial Theorem, we expand to the required number of terms the square root term in equation (3.31)

$$\sqrt{1-\frac{\sin^2\theta}{n^2}}=1-\frac{\sin^2\theta}{2n^2}-\frac{\sin^4\theta}{8n^4}-\frac{\sin^6\theta}{16n^6}-\cdots \tag{3.34}$$

Therefore

$$x=r\left[\cos\theta+n\left(1-\frac{\sin^2\theta}{2n^2}-\frac{\sin^4\theta}{8n^4}-\frac{\sin^6\theta}{16n^6}-\cdots\right)\right] \tag{3.35}$$

Differentiating the above equation (3.35), we get the velocity and acceleration of the piston.

$$v=-\omega r\,(A_1\sin\theta+A_2\sin 2\theta+A_4\sin 4\theta+A_6\sin 6\theta+\cdots) \tag{3.36}$$

$$a=-\omega^2 r\,(A_1\cos\theta+2A_2\cos 2\theta+4A_4\cos 4\theta+6A_6\cos 6\theta+\cdots) \tag{3.37}$$

where

$$\begin{aligned}
A_1 &= 1\\
A_2 &= \frac{1}{2n}+\frac{1}{8n^3}+\frac{15}{256n^5}+\frac{35}{1024n^7}+\cdots\\
A_4 &= -\frac{1}{16n^3}-\frac{3}{64n^5}-\cdots\\
A_6 &= \frac{3}{256n^5}+\frac{15}{1024n^7}-\cdots
\end{aligned} \tag{3.38}$$

Other than the first harmonic, we find that both velocity and acceleration have even higher harmonics of the crank speed. The magnitudes of harmonics are also given in series form. We find that these series are highly converging with n in the denominator. From an engineering stand-point, we can neglect all higher harmonics other than the second harmonic.

These harmonics can be used in developing primary and secondary forces and their balancing; we will study these later in designing a good multi-cylinder engine. These engines are ubiquitous in automobiles, Diesel Electric Railways, Power Plants, Marine Propulsion Systems and they form what we call *Drive Trains*. These drive trains will contain several gear boxes, couplings, propeller shafts, etc., in the system and they are all subjected to the above harmonic fluctuations. Therefore the acceleration analysis presented in this chapter forms the backbone of Theory of Machines and Design.

3.6 Additional Problems

1. Figure 3.47 shows a six-bar mechanism. Crank 2 is driven at 100 RPM in clockwise direction. Determine all the centros. Find the velocity of the slider by the

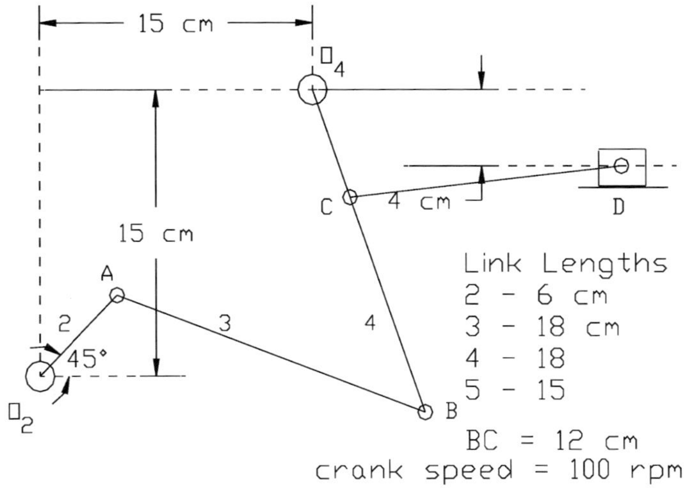

Fig. 3.47

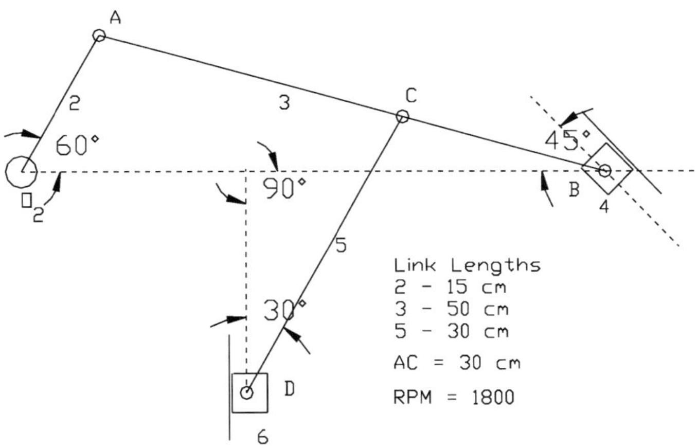

Fig. 3.48

Line of Centers method and show that the value obtained is the same as that obtained by the Link to Link method.

2. Figure 3.48 shows a six-bar mechanism with two sliders. The crank is driven at 1800 RPM in clockwise direction. Find the velocities of the two sliders by the method of centros.
3. Crank 2 in Figure 3.49 rotates at a speed of 60 RPM. Determine the velocity of slider 5 by the method of centros. What is the total stroke of this slider?
4. A six-bar linkage is shown in Figure 3.50. Determine the velocity of link 6 by the Method of Centros, Link to Link method and Line of Centers method.

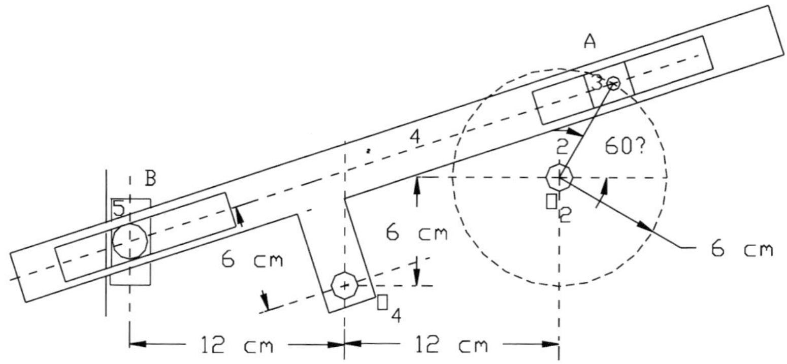

Fig. 3.49

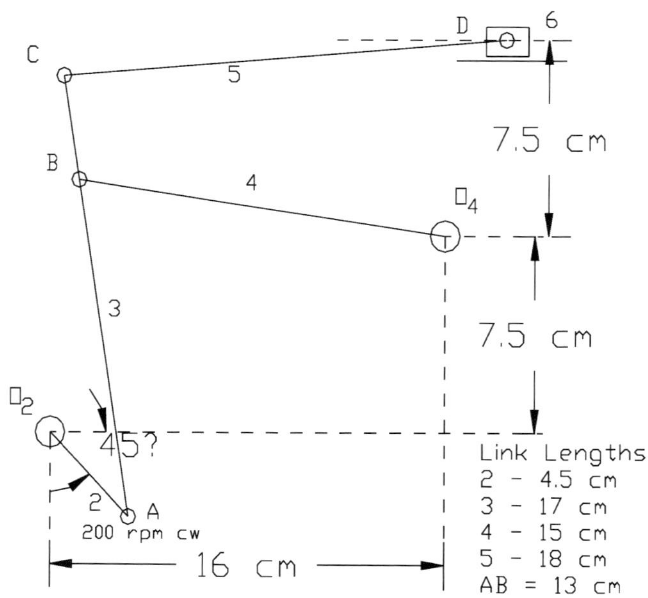

Fig. 3.50

5. What are the angular velocities of links 5, 6 and 7 in the eight-bar linkage of Figure 3.51? Also determine the velocities of both the sliders, links 4 and 8.
6. For the oscillating cylinder mechanism in Figure 3.52, locate all the centros. What is the angular velocity of the oscillating cylinder for the instant shown? Also determine the linear velocity of the piston.
7. Draw the linkage shown in Figure 3.53 and locate all the centros. Determine the angular velocities of links 5 and 6 by the method of centros for the crank rotating at 100 rad/s in counter clockwise direction.
8. Using the relative velocity approach, solve Additional Problem 1 above.

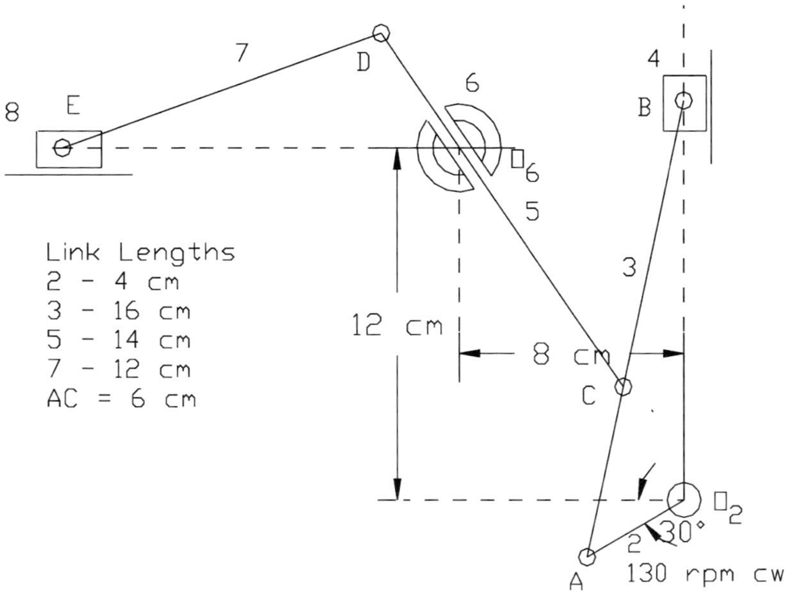

Fig. 3.51

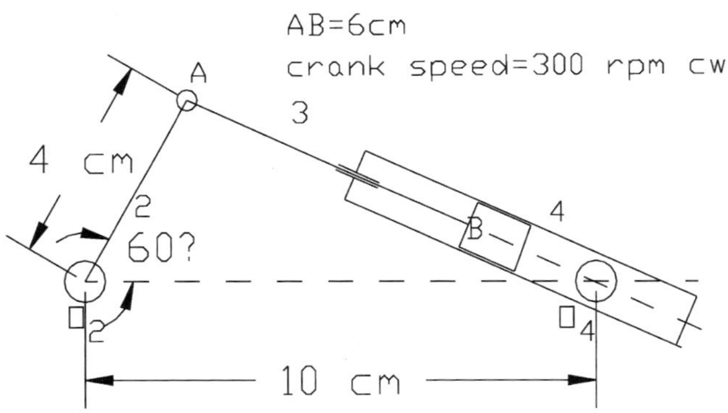

Fig. 3.52

9. Using the relative velocity approach, solve Additional Problem 2 above.
10. Using the relative velocity approach, solve Additional Problem 3 above.
11. Using the relative velocity approach, solve Additional Problem 4 above.
12. Using the relative velocity approach, solve Additional Problem 5 above.
13. Using the relative velocity approach, solve Additional Problem 6 above.
14. Using the relative velocity approach, solve Additional Problem 7 above.
15. Make a complete acceleration analysis of the mechanism in Figure 3.47.
16. Determine the accelerations of the two pistons and the angular acceleration of link 5 in Figure 3.48.

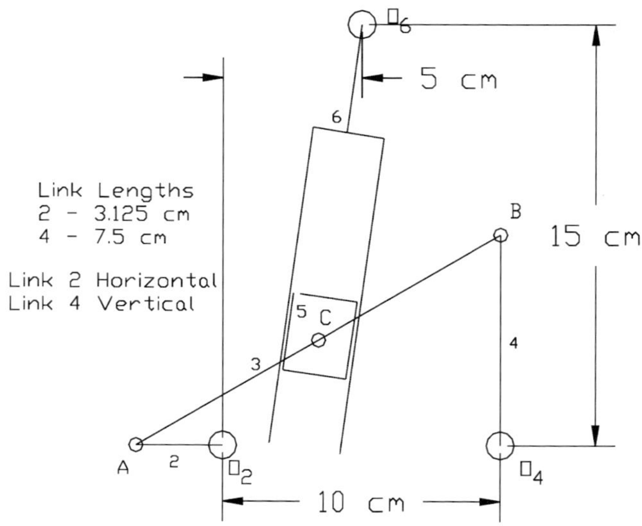

Fig. 3.53

17. Make a complete acceleration analysis of the linkage in Figure 3.49. What are the Coriolis components of accelerations for links 3 and 5 (including the directions)?
18. Determine the angular accelerations of links 3, 4 and 5 and the acceleration of the slider in Figure 3.50.
19. Refer to Figure 3.51. Identify the Coriolis component of acceleration and find the same in magnitude and sense. Determine the angular accelerations of links 3 to 7 and the linear accelerations of links 4 and 8.
20. Find the angular accelerations of links 3 to 6 in Figure 3.53.

Chapter 4
Straight Line Motion and Universal Coupling

Every link in a mechanism performs a repeated motion. Therefore any point on these links also performs a repeated motion passing through a specified curve called a coupler curve. In this chapter we will study special cases where the path traced is a straight line or special cases when two inclined shafts are connected through a universal coupling.

Coupler Curve

Curve traced by a point on one of the coupler links in a mechanism.

Exact Straight Line Motion Mechanism

Mechanism containing a link in which a point describes a coupler curve as an exact straight line.

Approximate Straight Line Motion Mechanism

Mechanism containing a link in which a point describes a coupler curve having a portion of it as an approximate straight line.

The interest in straight line or approximate straight line motion became important with the advent of beam engines such as those shown in Figure 4.13. Producing an exact straight line is difficult and hence approximate straight line motion was attempted. Watt's approximate straight line motion played a significant role in the initial designs of the reciprocating steam engine.

With further developments, engineers began to develop testing of their engines and recording of the data. The initial instruments developed were Indicators and a pen to record the motion. These indicators related some state quantities of the engine and utilized the principles of straight line motions. Other applications were

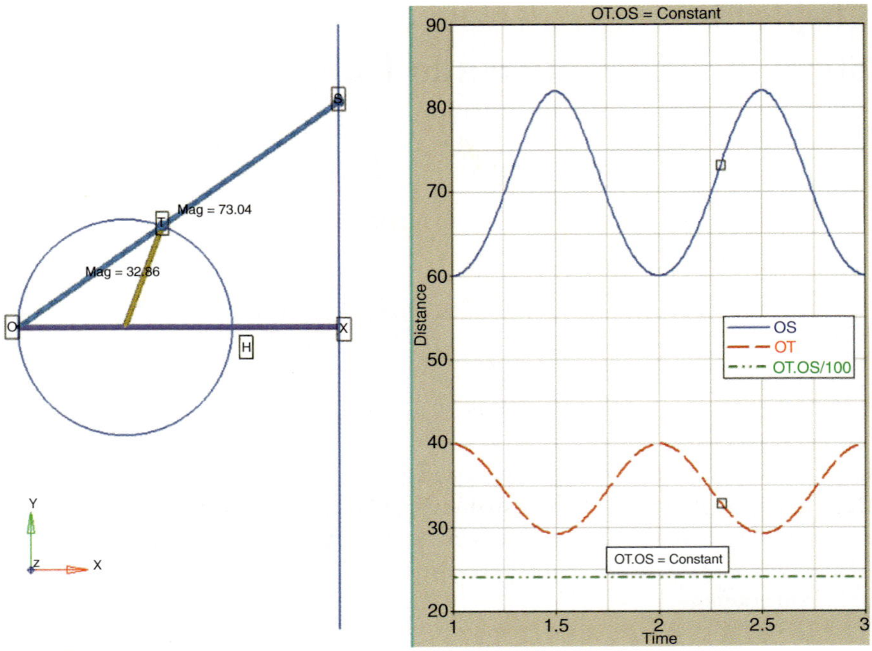

Fig. 4.1 Condition to be satisfied to generate an exact straight line

e.g., steering gears designed for steering an automobile around turns. We describe these linkages below.

4.1 Condition for Exact Straight Line Motion

We will first study the conditions to be satisfied to generate an exact straight line. Let O, T and S be three distinct points of a mechanism, which lie in a straight line, for all configurations as shown in Figure 4.1.

The path of S will be a straight line perpendicular to the horizontal diameter OH of the circle along the circumference of which T moves, provided $OT.OS$ is a constant. It follows from similar triangles OTH and OSX that

$$OX = \frac{OS \times OT}{OH} \tag{4.1}$$

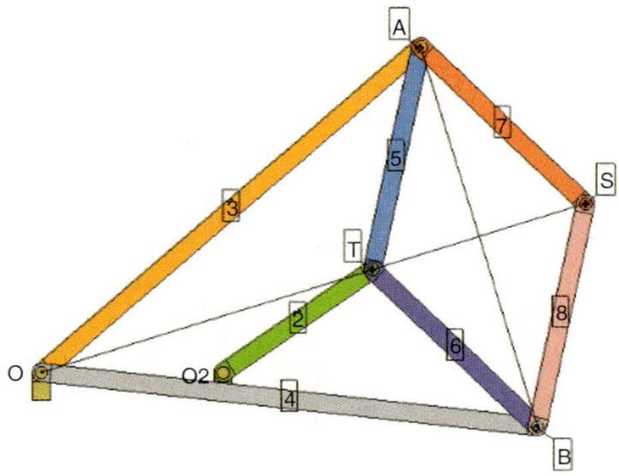

Fig. 4.2 Paucellier mechanism

4.2 Exact Straight Line Motion Mechanisms

4.2.1 Paucellier Mechanism

Figure 4.2 shows this linkage in which link 2 is the driver. Links 5 to 8 are of equal length so also are links 3 and 4. S traces an exact straight line perpendicular to the base link 1 going through OO_2. It can be easily shown that condition (4.1) is satisfied from the triangles *OAC* and *ACS*

$$\begin{aligned} OA^2 &= AC^2 + OC^2 \\ AS^2 &= CS^2 + AC^2 \end{aligned} \tag{4.2}$$

Subtracting one from the other,

$$OA^2 - AS^2 = OC^2 - CS^2 = OT \times OS \tag{4.3}$$

4.2.2 Hart Mechanism

This mechanism has links 3 to 6 forming a crossed parallelogram as shown in Figure 4.3. Crank 2 is the driving link connecting the base link 1 and crossed chain link 6 through the hinges O_2 and T such that $BO : BA = BT : BD$. Point S on link 5 traces an exact straight line.

From similar triangles *AOS* and *ABC*

$$BC = BA \times \frac{OS}{OA} \tag{4.4}$$

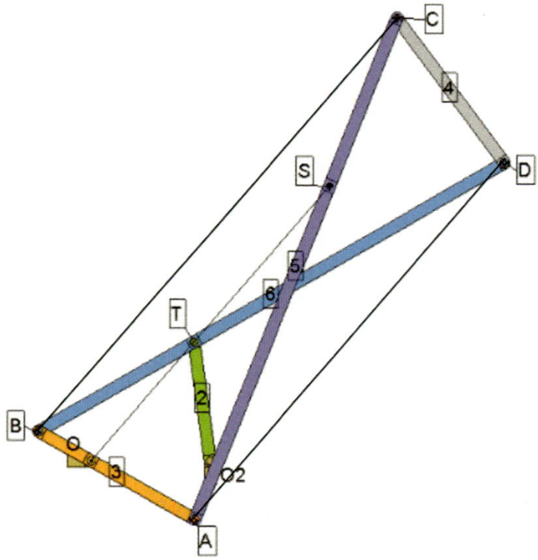

Fig. 4.3 Hart mechanism

Similarly, from similar triangles *BOT* and *BAD*,

$$AD = BA \times \frac{OT}{OB} \tag{4.5}$$

Therefore,

$$OS \times OT = K \times BC \times AD \tag{4.6}$$

From triangle *ACD*,

$$AC^2 = AD^2 + CD^2 - 2 \times AD \times CD \times \cos A\hat{D}C \tag{4.7}$$

From Figure 4.3

$$\cos A\hat{D}C = \frac{BC - AD}{2 \times CD} \tag{4.8}$$

Therefore

$$\begin{aligned} AC^2 &= AD^2 + CD^2 - AD\,(AD - BC) \\ &= CD^2 + BC \times AD \end{aligned} \tag{4.9}$$

Hence $BC \times AD$, therefore $OS \times OT$ is a constant.

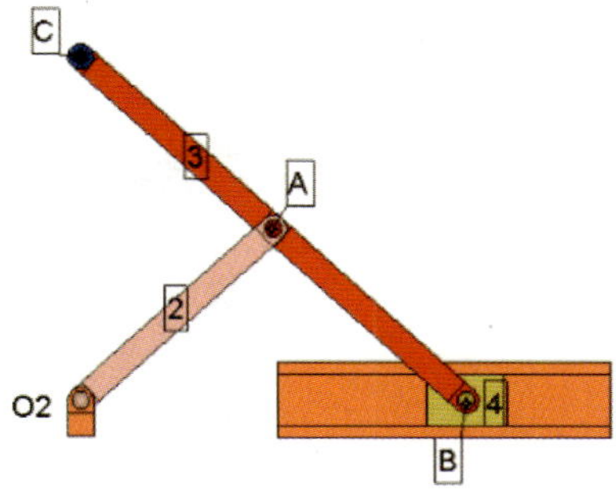

Fig. 4.4 Scott–Russel mechanism

4.2.3 Scott–Russel Mechanism

This mechanism has just four links as shown in Figure 4.4 with the condition $O_2A = AB = AC$.

4.3 Approximate Straight Line Motion Mechanisms

In several cases we use only approximate straight line motion, which will suffice, e.g., James Watt in his reciprocating engine used approximate straight line motion for the piston.

4.3.1 Modified Scott–Russel (Grasshopper) Mechanism

Here, the slider 4 in Figure 4.4 is replaced by a rocker link 4. The two extreme positions of B are B and B_1 as shown in Figure 4.5 corresponding to the positions A or A' and A_1 respectively. The path traced by C on coupler 3 passes through C, O_2 and C' and will approach a straight line if link 4 is sufficiently longer than the other three links.

This mechanism can be further modified as shown in Figure 4.6, so that C traces a path which is offset from the crank center O_2. The links are proportioned $O_2A/AB = AB/AC$ so that I is the instantaneous center of link 3 and C will describe a vertical line in this position.

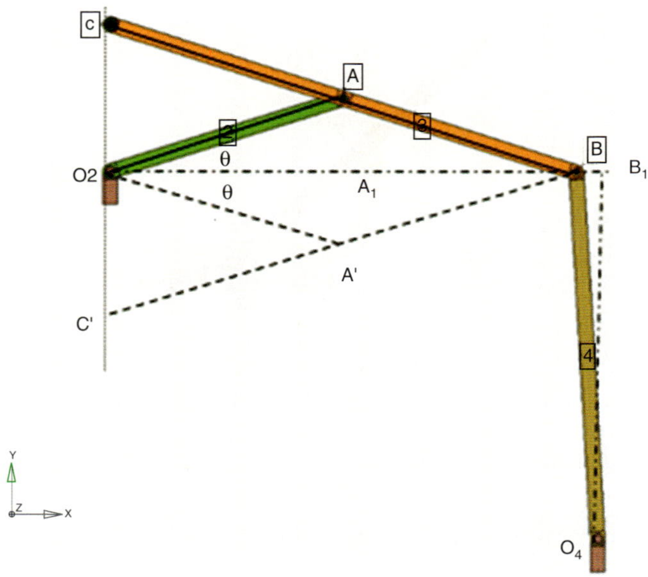

Fig. 4.5 Modified Scott–Russel (grasshopper) mechanism

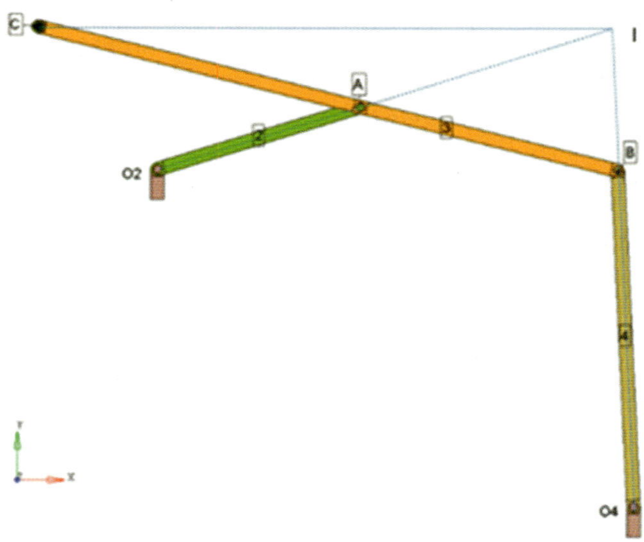

Fig. 4.6 Further modified Scott–Russel mechanism

4.3.2 *Watt Mechanism*

This is a four-bar linkage so arranged that point C on coupler link 3 describes an approximate straight line. When rockers 2 and 3 are in horizontal position, the coupler is made to be vertical, see Figure 4.7. The best position for C is determined

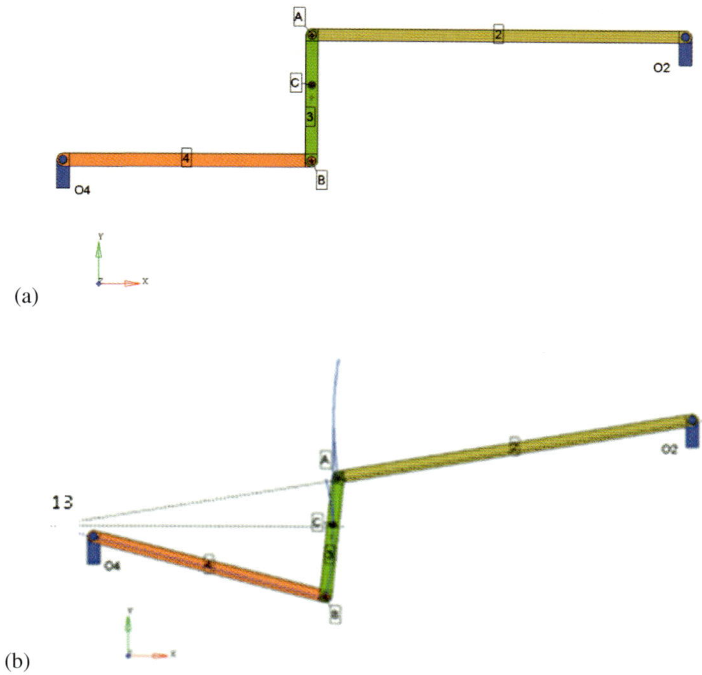

Fig. 4.7 (a) Watt mechanism; (b) Watt mechanism in displaced position

by using the concept of instantaneous center 13 of coupler 3 in a displaced position with the coupler in $A'B'$ position ($'$ positions refer to Figure 4.7b). In this position, point C' and the whole coupler will have a vertical motion. For small values of crank rotations

$$\frac{A'C'}{C'B'} = \frac{O_4B'}{O_2A'} \tag{4.10}$$

Therefore,

$$\frac{AC}{CB} = \frac{O_4B}{O_2A} \tag{4.11}$$

4.3.3 Tchebicheff Mechanism

As shown in Figure 4.8 this mechanism has its midpoint P on a coupler describing an approximate straight line P_2PP_1. The proportions of the links are in the ratio

$$O_2O_4 : (O_2A = O_4B) : AB = 4 : 5 : 2 \tag{4.12}$$

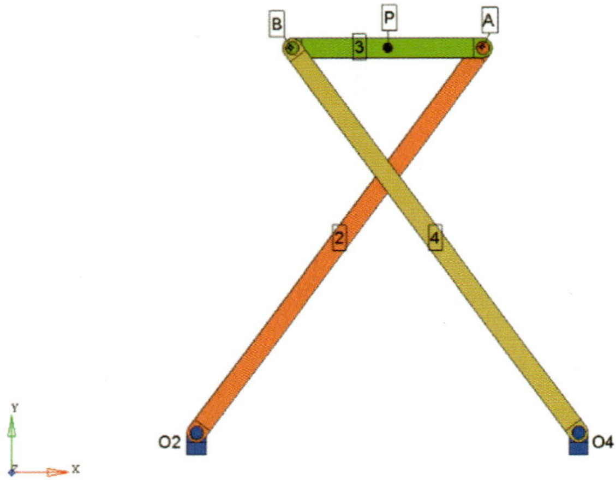

Fig. 4.8 Tchebicheff mechanism

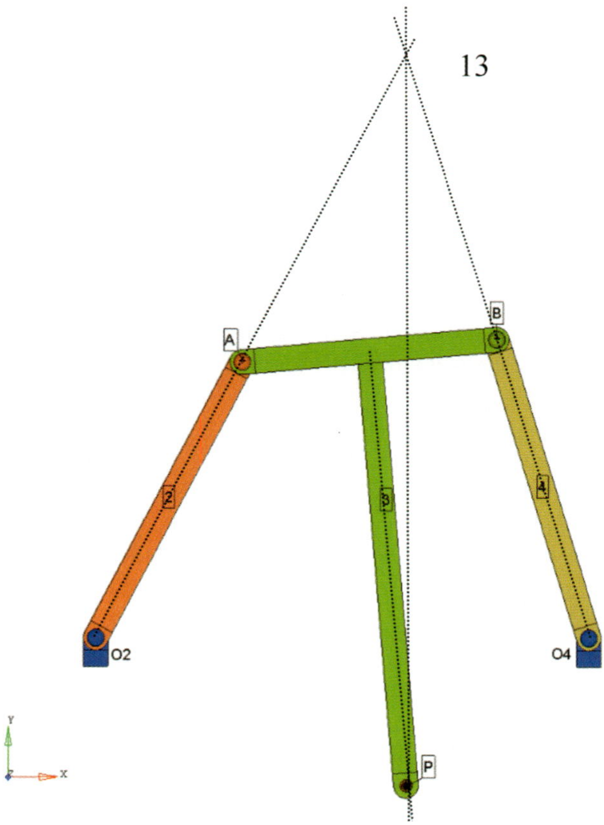

Fig. 4.9 Robert straight line mechanism

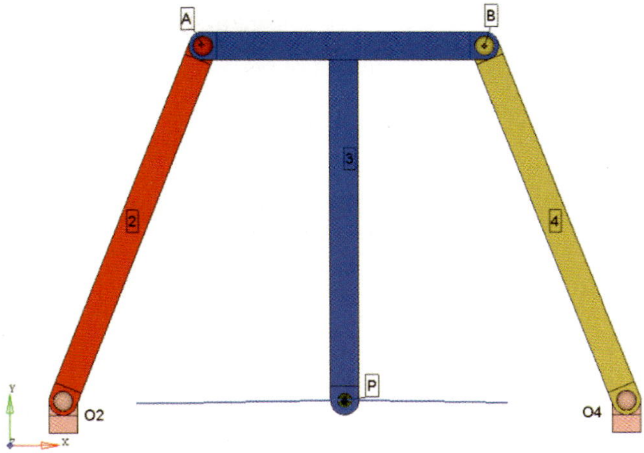

Fig. 4.10 Special case of Robert straight line mechanism

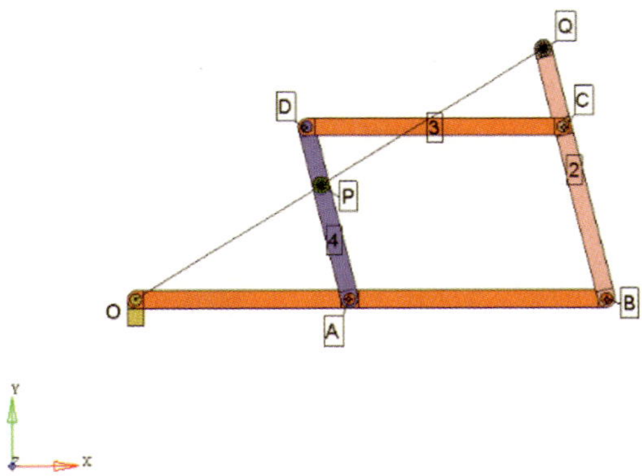

Fig. 4.11 Pantograph

4.3.4 Robert Straight Line Mechanism

In this linkage shown in Figure 4.9, the rockers are of equal length. Choose an instantaneous position and locate the instantaneous center (13). P is on the perpendicular bisector of the coupler intersecting with the vertical line dropped from the instantaneous center (13) of the coupler.

Figure 4.10 shows a special case of this linkage in which $2 \times AB = O_2O_4$.

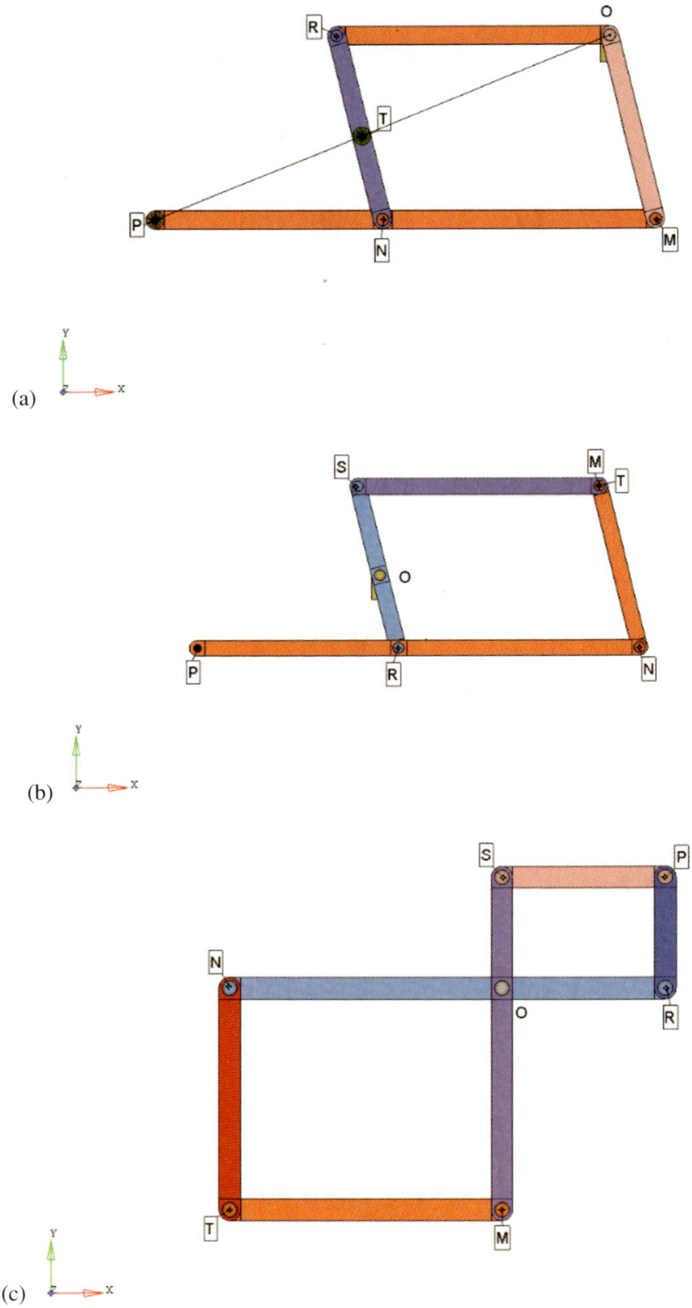

Fig. 4.12 Pantographs

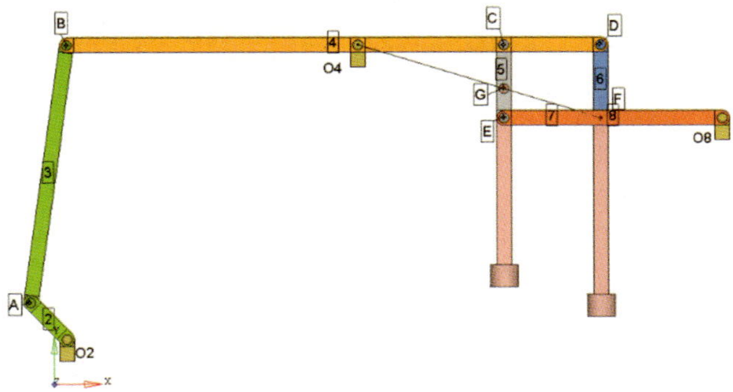

Fig. 4.13 Beam engine mechanism

4.3.5 Pantograph

A pantograph has an arrangement of links as shown in Figure 4.11. Link 1 is hinged at O. AD is parallel to BC for all positions of the linkage.

$$\frac{OP}{OQ} = \frac{OA}{OB} \tag{4.13}$$

Point Q traces a path similar to that of point P and vice versa. Figure 4.12 shows three other arrangements of the pantograph.

4.3.6 Beam Engine

This mechanism was used by James Watt in his steam engine. This is an eight link mechanism as shown in Figure 4.13. The steam piston is connected to the hinge F, while the pump piston is connected to a hinge G on link 5. Links 4 to 7 form a pantograph giving similar motions for G and F. Links 1, 4, 5 and 8 form Watt's linkage for approximate straight line motion for G and F. Hinge G is centrally located on link 5 between E and C. O_4C is made equal to O_8E.

4.3.7 Richards Indicator

Here also, the Watt linkage is used in conjunction with a pantograph as illustrated in Figure 4.14.

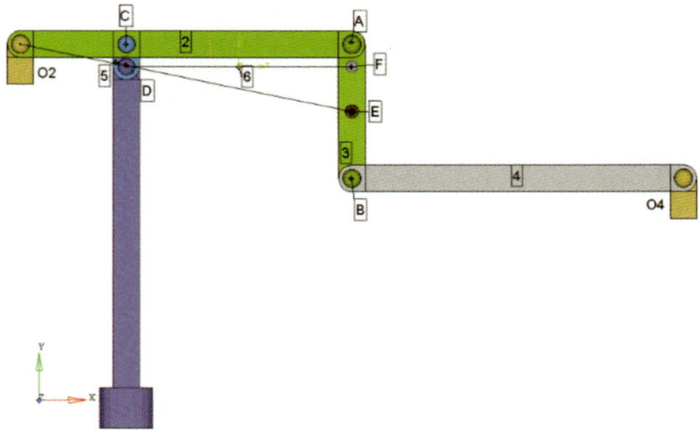

Fig. 4.14 Richards indicator

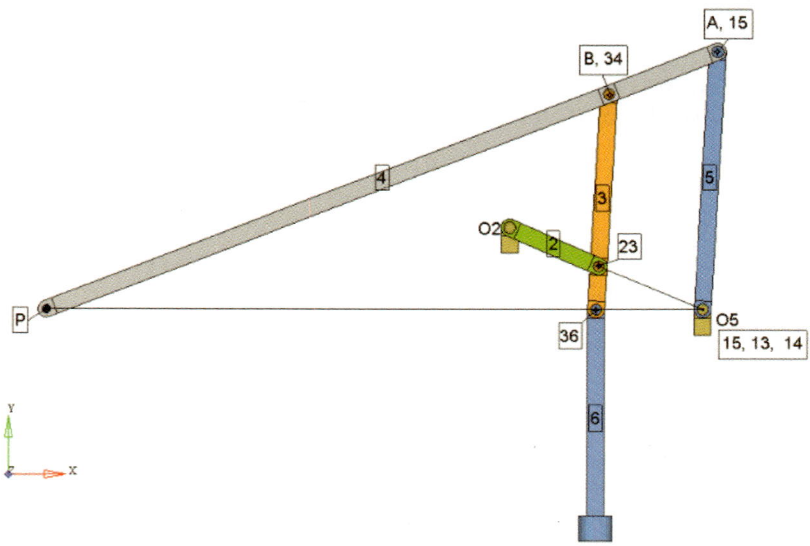

Fig. 4.15 Crosby indicator

4.3.8 Crosby Indicator

This is a six-link mechanism as shown in Figure 4.15. For the configuration shown, it can be seen that O_5 is the instantaneous center of links 3 to 5. (36) is the centro of links 3 and 6. Choose the point P on link 4 such that P (36) and O_5 lie on a horizontal line.

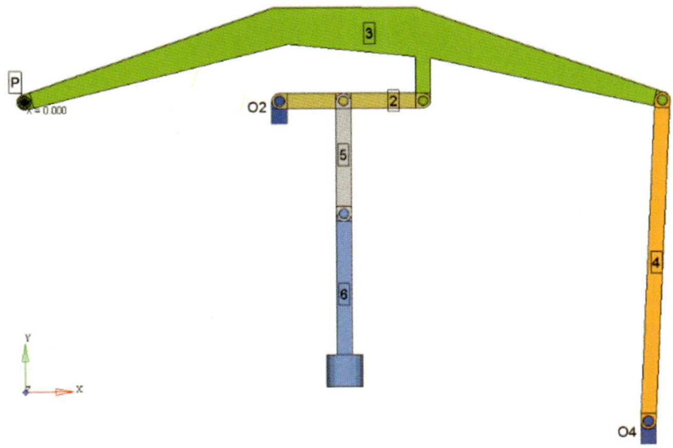

Fig. 4.16 Dobbie–McInnes mechanism

4.3.9 Dobbie–McInnes Mechanism

Figure 4.16 shows this linkage. Point P on link 3 forming a horizontal line with instantaneous center (13) describes a vertical line for the configuration shown. (One can show that (13) is the same as centro (34) by considering links 1, 2, 3 and 1, 3, 4.)

4.4 Steering Gear Mechanism

The condition for correcting steering of all wheels in an automobile is to make them turn about a single instantaneous center of rotation I as shown in Figure 4.17. Then

$$AC = EF = EI = FI = AE\cot\phi - CF\cot\theta$$

$$\cot\phi - \cot\theta = \frac{a}{b} \tag{4.14}$$

4.4.1 Davis Steering Gear Mechanism

This linkage is shown in Figure 4.18a where the arms AE and CF are fixed to the axles in such a way as to form bell-crank levers BAE and DCF with equal angles. These arms are also slotted and slide relative to the two die-blocks pivoted to the link GH. Link GH is made to move parallel to AC at a distance h and it activates

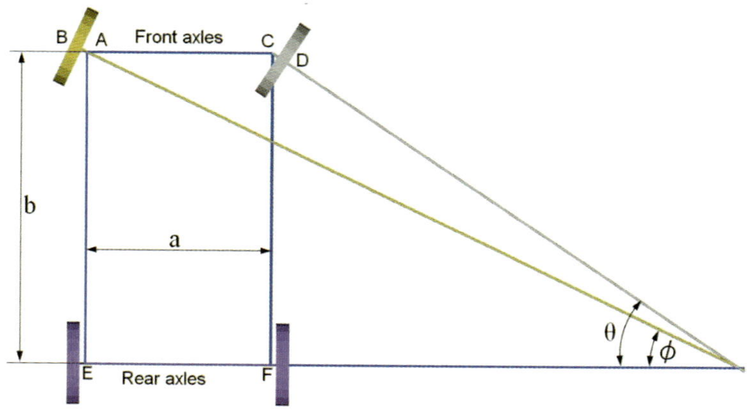

Fig. 4.17 Condition for correct steering linkage

steering to the right or left. In the neutral position, AE and CH make an angle α as shown in Figure 4.18a.

When GH is displaced by a distance x, the linkage occupies the position as shown. BA and CD in this position meet at I (not shown in figure). Let ϕ and θ be the angles through which the arms AE and CF are turned by the displacement x of GH. Let the difference between AC andGH be $2c$, then

$$\tan(\alpha + \phi) = \frac{\tan\alpha + \tan\phi}{1 - \tan\alpha\tan\phi} = \frac{c + x}{h} \quad \text{and} \quad \tan\alpha = \frac{c}{h},$$

and we get

$$\tan\phi = \frac{xh}{h^2 + c^2 + cx} \tag{4.15}$$

Similarly,

$$\tan(\alpha - \theta) = \frac{\tan\alpha - \tan\theta}{1 + \tan\alpha\tan\theta} = \frac{c - x}{h}$$

and this gives

$$\tan\theta = \frac{xh}{h^2 + c^2 - cx} \tag{4.16}$$

Therefore

$$\cot\phi - \cot\theta = 2\tan\alpha \tag{4.17}$$

From equation (4.14) we can now get the condition for correct steering in the Davis linkage

$$\tan\alpha = \frac{a}{2b} \tag{4.18}$$

Figure 4.18b shows the linkage designed to satisfy the above relation. In the accompanying plot you are given the left and right wheel angles as they change with time during a right turn. The dash-dot line shows the relation (4.17) which

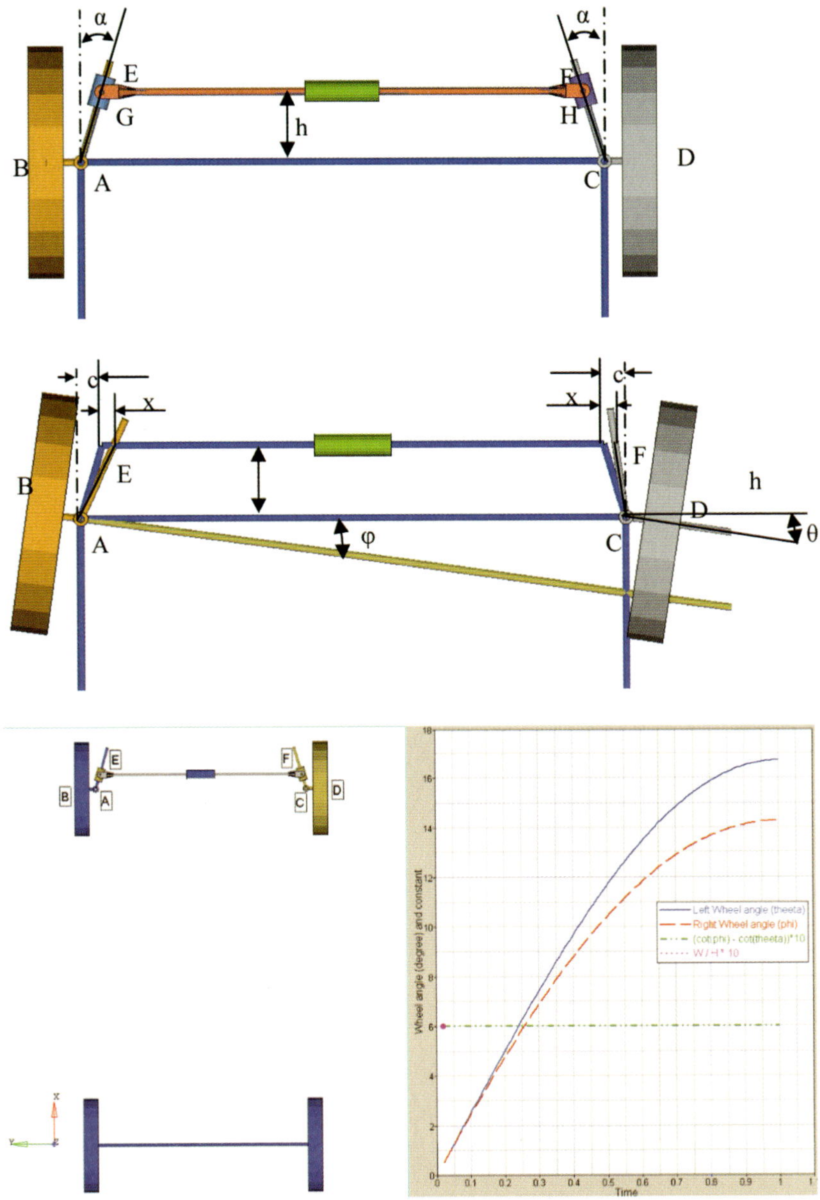

(a)

(b)

Fig. 4.18 Davis steering linkages

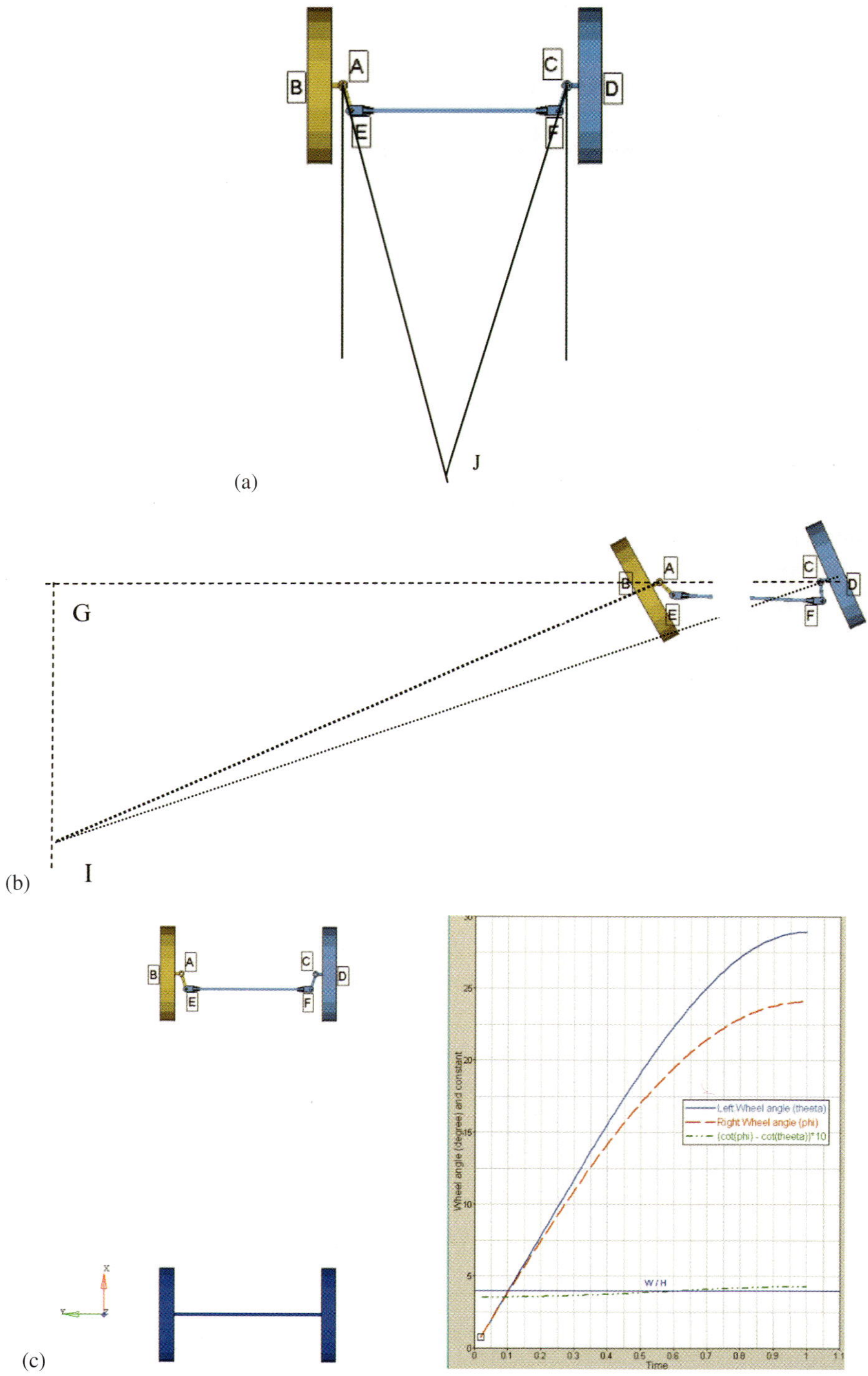

Fig. 4.19 (a) Ackermann linkage, (b) Ackermann linkage after a left turn, (c) Ackermann steering gear

remains the same throughout the turning process, thus giving exact steering to the car.

4.4.2 Ackermann Steering Gear Mechanism

This is a four-bar linkage with links $AE = CF$, AC and EF. BAE and DCF form bell-crank levers. The neutral position with links AE and CF making an angle α as well as the position in one of the displaced configurations in a left turn condition are shown in Figures 4.19a and b.

Figures 4.19b and c give MotionSolve solutions for the left and right turns during steering respectively.

It can be seen that $\cot\phi - \cot\theta = AC/GI$ where I is the point of intersection of BA and CD. This is obviously not a constant for all steering positions and therefore cannot satisfy the condition (4.14). The advantage of this linkage is that it has no sliding pairs and therefore is easy to maintain without much wear and tear (thus not introducing errors in the linkage positions). The best possible steering is found when the point J from AC is at about 0.7 times the wheel base.

4.5 Hooke's (Cardan, Universal) Joint or [Universal Coupling]

This is a kinematic joint connecting two shafts with intersecting axes, see Figure 4.20. It consists of three parts, two forks (driving and driven shafts) and a center block or spider. This is actually a space linkage, which however, can be analyzed in a simple manner.

The plan and elevation of the joint is shown in Figure 4.21 with δ representing the inclination angle between the two shafts. The fork is represented by lines aa and bb. When the driving shaft is rotated aa describes a circle, whereas the driven shaft bb describes an ellipse as shown. For an angle of rotation of the input shaft α, aa moves to a_1a_1, the projection of bb also moves through the same angle to b_1b_1. The true position of bb is obtained from projecting b_1 into the plan resulting in b_1', which again projected into the elevation gives b_2.

Then

$$\tan\alpha = \frac{Oc_1}{b_1c_1}$$

$$\tan\beta = \frac{Oc_2}{b_2c_2} = \frac{Oc_2}{b_1c_1}$$

Also, from the plan view,

$$\cos\delta = \frac{Oc_1}{Ob_1{}'} = \frac{Oc_1}{Oc_2}$$

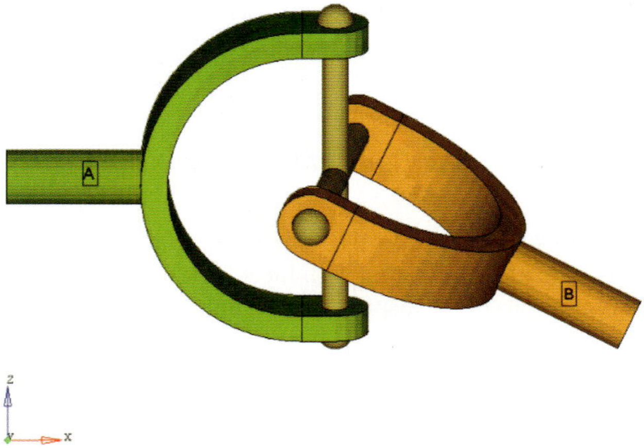

Fig. 4.20 Hooke's joint

Therefore

$$\tan\alpha = \tan\beta\cos\delta \tag{4.19}$$

Differentiating the above,

$$\omega_a \sec^2\alpha = \omega_b \sec^2\beta\cos\delta = \omega_b(1+\tan^2\beta)\cos\delta \tag{4.20}$$

With the help of (4.19) and rearranging of equation (4.20), we get

$$\frac{\omega_b}{\omega_a} = \frac{\cos\delta}{1-\sin^2\delta\cos^2\alpha} \tag{4.21}$$

When $\cos\alpha = \pm 1(\alpha = 0, 180°)$, the above ratio is maximum given by $1/\cos\delta$ and when $\cos\alpha = 0(\alpha = 90, 270°)$, the speed ratio is minimum given by $\cos\delta$. The coefficient of fluctuation of speed is therefore

$$\frac{\frac{1}{\cos\delta} - \cos\delta}{1} = \sin\delta\tan\delta \tag{4.22}$$

From (4.21), we find that the driver and driven shafts have the same speed when

$$\frac{\cos\delta}{1-\sin^2\delta\cos^2\alpha} = 1$$

Upon simplification, we get

$$\tan\alpha = \pm\sqrt{\cos\delta} \tag{4.23}$$

For a constant driving speed, $\dot{\omega}_a = 0$ the acceleration of the driven shaft from equation (4.21) can be obtained as

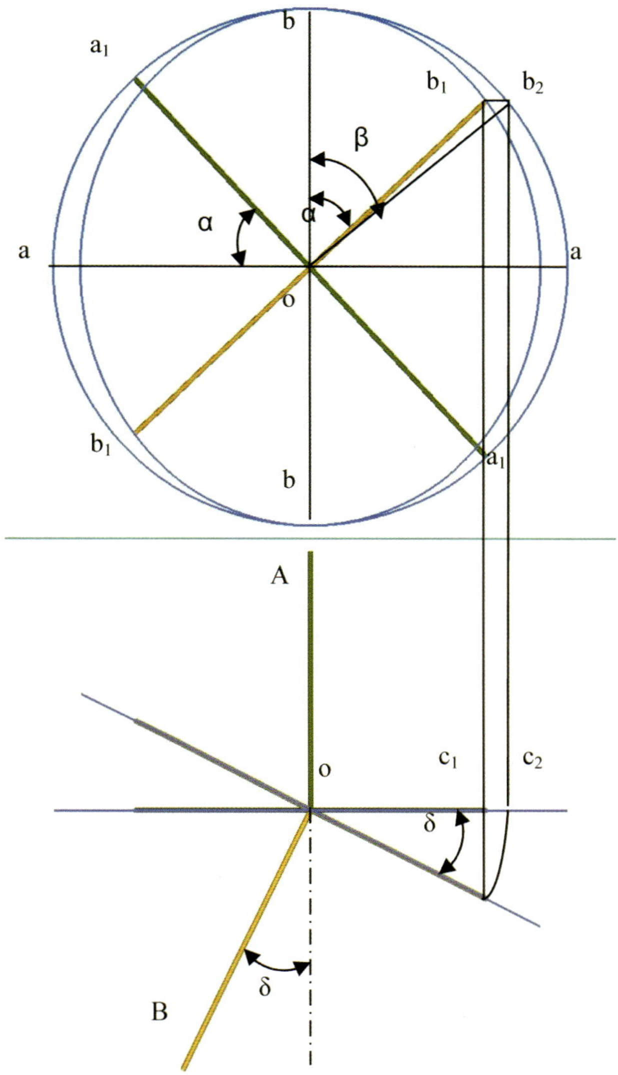

Fig. 4.21 Analysis of Hooke's joint

$$\dot{\omega}_b = -{\omega_a}^2 \left[\frac{\cos\delta \sin^2\delta \sin 2\alpha}{\left(1 - \sin^2\delta \cos^2\alpha\right)^2} \right] \tag{4.24}$$

Differentiating the above with respect to α, we get the acceleration of the driven shaft to be maximum at

$$\cos 2\alpha = \frac{\sin^2\delta(2 - \cos^2 2\alpha)}{2 - \sin^2\delta} \approx \frac{2\sin^2\delta}{2 - \sin^2\delta} \tag{4.25}$$

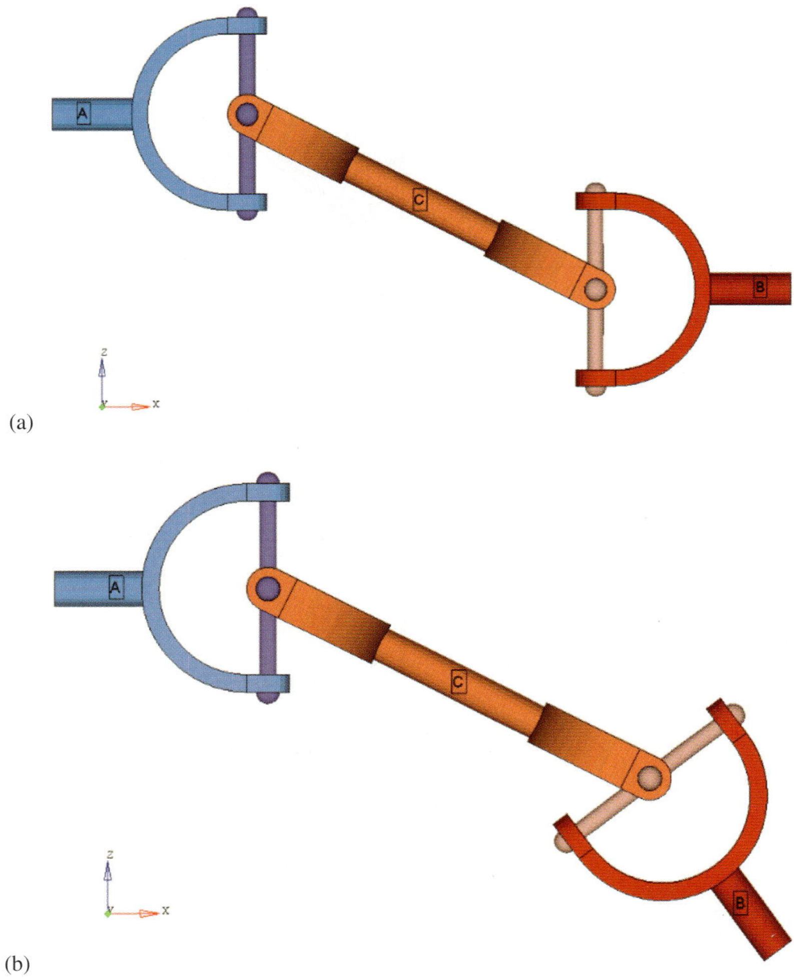

Fig. 4.22 Double Hooke's joint intermediate shaft forks in the same plane: (a) Arrangement a. (b) Arrangement b

4.5.1 Double Hooke's Joint

This type of coupling is used to transmit motion between two parallel shafts without using gears, belts or chains. Figures 4.22a, b and c illustrate two different ways of doing this.

In Figures 4.22a and b, the two forks at the end of the intermediate shaft are in the same plane and the inclination angles are the same at both ends. From equation (4.19)

$$\tan\alpha = \tan\gamma\cos\delta \tag{4.26}$$

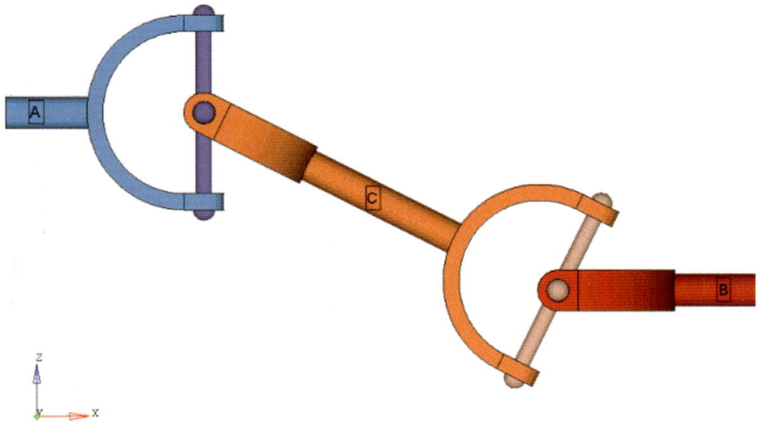

(c)

Fig. 4.22 (Continued) (c) Arrangement c

where γ is the angle turned by the intermediate shaft C. Similarly,

$$\tan\beta = \tan\gamma\cos\delta \tag{4.27}$$

Therefore, $\alpha = \beta$ and in this case, both the shafts turn through the same angle giving

$$\omega_b = \omega_a \tag{4.28}$$

If the two end forks of the shaft C are set at right angles as in Figure 4.22c, then equation (4.27) is

$$\tan\gamma = \tan\beta\cos\delta \tag{4.27a}$$

Combining equations (4.26) and (4.27a), we get $\tan\alpha = \tan\beta\cos^2\delta$. Differentiating this relation, we get

$$\frac{\omega_b}{\omega_a} = \frac{1}{\cos^2\delta\cos^2\alpha + \frac{\sin^2\alpha}{\cos^2\delta}} \tag{4.28a}$$

Therefore, the output shaft speed is not constant, instead it varies between $\omega_a\cos^2\delta$ and $\omega_a/\cos^2\delta$.

4.6 Solved Problems

Solved Problem 4.1

The dimensions of a Watt mechanism are $O_2O_4 = 65$ cm, $O_2A = O_4B = 30$ cm, $QA = QB = 7.5$ cm. Trace the locus of the coupler point Q.

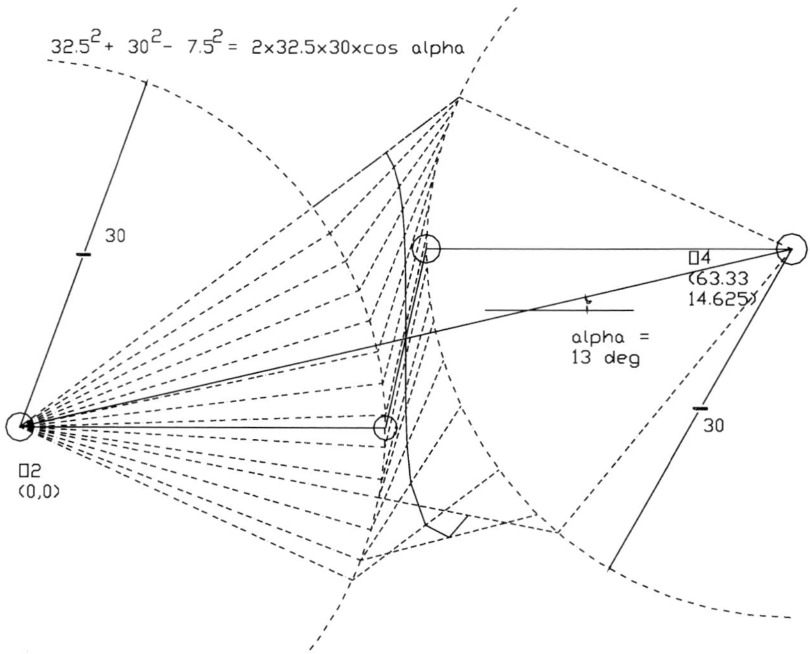

Fig. 4.23

Figure 4.23 shows the linkage. The linkage is drawn when the two levers are in a horizontal position so that the coupler curve is approximately a vertical line. This position can be determined from the geometry given in Figure 4.23, viz., $32.5^2 + 30^2 - 7.5^2 = 2 \times 32.5 \times 30 \times \cos\alpha$ giving 13.0028°.

Solved Problem 4.2

Roberts's straight line mechanism of Figure 4.10 has the following proportions: $O_2B = BP = PC = CO_4$ and BC is half of the base link. Trace the locus of the coupler point P for $o \leq O_4\hat{O}_2B \leq 75°$ (see Figure 4.24).

Solved Problem 4.3

For the Freemantle mechanism shown in Figure 4.25, show that the locus of the point M is an exact straight line. The lengths $OA = AB = MA$.

Draw the linkage in any arbitrary position and a circle going through the three points O, B and M. We find the sum of the angles to be $p + q + (p + q) = 180°$. Therefore the sum of angles p and q at O is 90°. Therefore *OM* remains vertical in all positions of the mechanism. The above result can also be obtained by observing

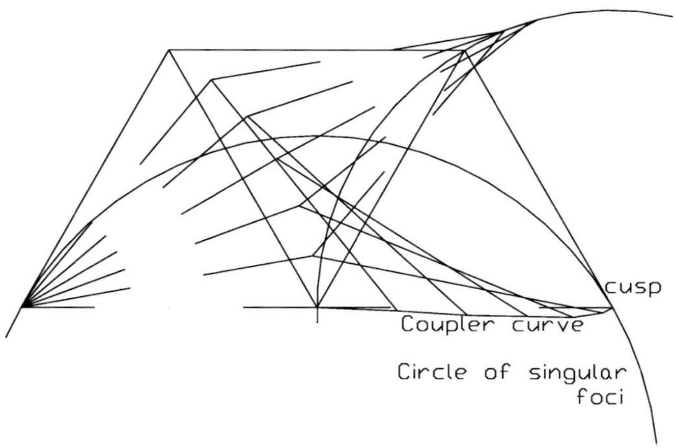

Fig. 4.24

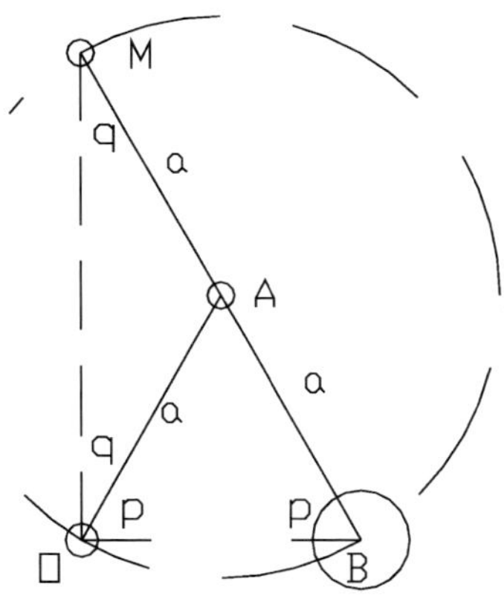

Fig. 4.25

that points M, O and B are on a semicircle with *MB* as a diameter. Therefore the angle subtended at O is always 90°.

Solved Problem 4.4

Design a pantograph for an indicator to obtain the indicator diagram of an engine. The distance from the tracing point of the indicator is 100 mm. The indicator diagram should represent four times the gas pressure inside the cylinder of an engine.

Refer to Figure 4.11. We are given $OP/OQ = 1/4$ and $DQ = 100$ mm. Therefore $OP = 25$ mm. Also, $OA/OB = 1/4$. Choose *OA* to be 10 mm, then $OB = 40$ mm.

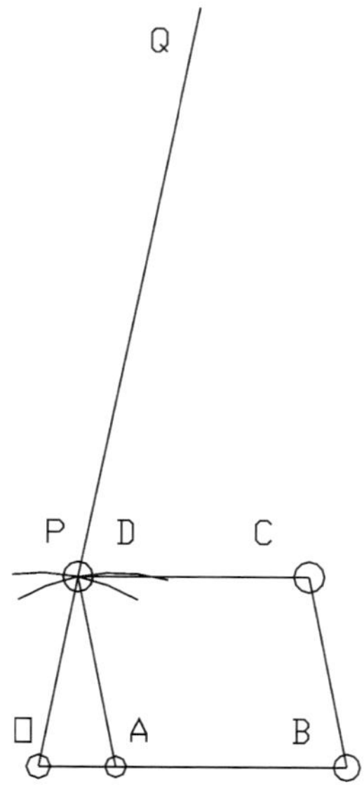

Fig. 4.26

Draw *OA* and *AB* as shown in Figure 4.26. Choose points *D* and *P* to be the same where the piston rod of the indicator is connected. Draw arcs 25 mm with *O* and *A* as centers to intersect at *P*. Extend *OP* to *Q* and complete the parallelogram *ABCD*.

Solved Problem 4.5

In a Davis steering gear mechanism, derive an expression for the condition of correct steering. Find the inclination of the track arms to the longitudinal axis of the car when the car is moving on a straight path. We are given that the distance between the pivots of the front axle is 1.0675 m and the wheel base is 2.5925 m.

For the derivation see Section 4.4.1. From equation (4.18)

$$\alpha = \tan^{-1} \frac{1.0675}{2 \times 2.5925} = 11.634^\circ$$

Solved Problem 4.6

An Ackermann steering gear mechanism is used for an automobile with a track of 1.475 m and a wheel base of 2.745 m, with a track rod in front of the axle instead of behind it. Determine the radius of curvature of the path followed by the near-side front wheel when the automobile turns right and assuming correct steering exists.

We are given that the length of the track rod is 1.334 m, the length of each track arm is 0.1525 m and the distance between the front stub axle pivots is 1.22 m.

1. The Ackermann steering gear for this problem is drawn in Figure 4.27. From this we find that
$$\sin\alpha = \frac{1.334 - 1.22}{2 \times 0.1525}$$
gives $\alpha = 21.9483°$.
2. Try various values of θ for the inner wheel and find the corresponding values of ϕ e.g., for $\theta = 31°$, $\phi = 24.9493°$. From (4.14), however,
$$\cot\phi = \cot 31° + \frac{1.22}{2.745} = 25.37°$$
Therefore correct steering does not exist for this turning angle. For $\theta = 33°$, $\phi = 27°$, for which equation (4.14) gives $\phi = 26.746°$. This position is taken as the correct steering position. The radius of curvature of the automobile path is then 5.2987 m as given in Figure 4.27.

Solved Problem 4.7

Two shafts are connected by a Hooke's joint; the axes of the shafts are inclined at 15° and the speed of the driving shaft is 800 RPM. Find the highest and lowest speeds of the driven shaft and its maximum acceleration.

Given $\omega_a = (2\pi \times 800)/60 = 83.7758$ rad/s.

1. From equation (4.21), we find that ω_b of the driven shaft is maximum for angles of the driven shaft α at 0°, 180°, etc., and has a value of $\omega_b = \omega_a/\cos\delta$. Therefore, the highest speed of the driven shaft is $\omega_{b\,\max} = 83.7758/\cos 15° = 86.7311$ rad/s.
2. From equation (4.21), we find that ω_b of the driven shaft is minimum for angles of the driven shaft α at 90°, 270°, etc., and has a value of $\omega_b = \omega_a \cos\delta$. Therefore, the lowest speed of the driven shaft is $\omega_{b\,\min} = 83.7758 \cos 15° = 80.9212$ rad/s.
3. Maximum acceleration for the driven shaft occurs at
$$\cos 2\alpha = \frac{2\sin^2 15°}{2 - \sin^2 15°} = 0.0693087$$
see equation (4.25), or when $\alpha = 43.0129$ and $223.0129°$.
4. Equation (4.24) gives the acceleration of the driven shaft and at the above values of α, it is

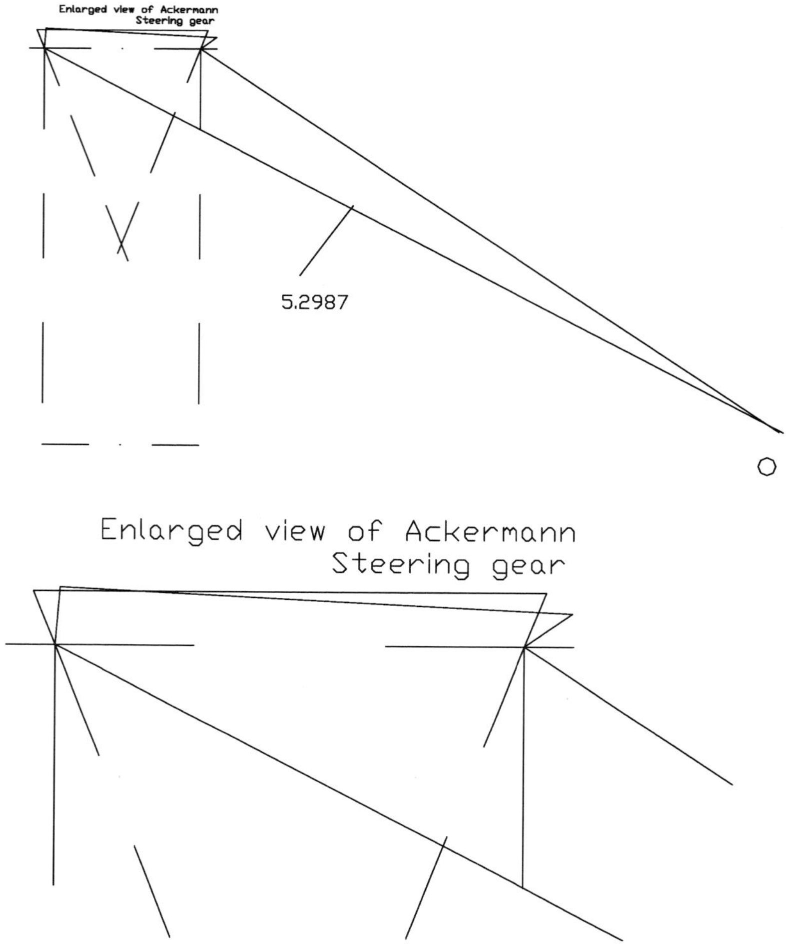

Fig. 4.27

$$\dot{\omega}_b = {\omega_a}^2 \left[\frac{\cos\delta \sin^2\delta \sin 2\alpha}{\left(1 - \sin^2\delta \cos^2\alpha\right)^2} \right] = -83.7758^2 \left[\frac{0.9659 \times 0.06699 \times 0.9976}{(1 - 0.06699 \times 0.53465)^2} \right]$$

$$= -487.3 \text{ rad/s}^2$$

Solved Problem 4.8

Two shafts are connected by a Hooke's joint, whose axes are inclined to each other at an angle δ, show that the instantaneous ratio of speeds between these shafts is given by $\sec^2\theta / \sec^2\phi \cos\delta$ where θ and ϕ are the angles through which the two halves of the Hooke's joint have turned respectively from some datum. Determine also the minimum and maximum values of this ratio when $\delta = 15°$.

Follow the derivation of equation (4.20)

$$\frac{\omega_{b\,\text{min}}}{\omega_a} = \cos 15° = 0.9659$$

$$\frac{\omega_{b\,\text{max}}}{\omega_a} = \frac{1}{\cos 15°} = 1.0353$$

Solved Problem 4.9

A Hooke's joint connects two shafts whose axes are inclined to each other by an acute angle δ. The driving shaft rotates uniformly and the total variation in speed of the driven shaft is not to exceed 10% of the mean speed. Determine the greatest possible inclination of the center lines of the shafts.

1. From equation (4.22), the coefficient of speed fluctuation is given by $(1/\cos\delta) - \cos\delta$.
2. Therefore $(1/\cos\delta) - \cos\delta = 0.1$.
3. $1 - \cos^2\delta - 0.1\cos\delta = 0$.
4. Hence

$$\cos\delta = \frac{1}{2}\left[-0.1 \pm \sqrt{0.01 + 4}\right] = 0.951249 \quad \text{or} \quad -1.05$$

5. Only the first value in the above is applicable, which gives the shaft angle being 17.964°.

Solved Problem 4.10

Two shafts are connected by a Hooke's joint and the speed of the driving shaft is 400 RPM. If the speed of the driven shaft should always be between 420 and 380 RPM, determine the greatest permissible angle between the two shafts. What are the actual maximum and minimum shaft speeds for such an angle.

1. The permissible coefficient of fluctuation in speed is $(420 - 380)/400 = 0.1$.
2. From equation (4.22), we have $(1/\cos\delta) - \cos\delta$ and, as in the solved problem 4.9, the shaft angle $\delta = 17.964°$.
3. From equation (4.21), the actual maximum speed attainable for this angle of inclination of the shafts is

$$\omega_{b\,\text{max}} = \frac{400}{\cos 17.964°} = 420.5 \text{ RPM}$$

4. Similarly, the minimum attainable speed is $\omega_{b\,\text{min}} = 400 \times \cos 17.964° =$ 380.5 RPM.

Solved Problem 4.11

In a single Hooke's joint where the angle between the shafts is θ, show that the ratio of the fluctuation of speed to the mean speed is $\sin\theta \tan\theta$ if the angular velocity of the driving shaft is constant and the axes of the pins in the joint intersect.

When a double coupling is used, explain what conditions are necessary for the driven shaft to have uniform angular velocity if that of the driving shaft is constant? In such a double coupling, the driving and driven shafts are parallel and the angle between each and the intermediate shaft is 20°. Find the maximum and minimum velocities of the driven shaft if the axis of the driving pin carried by the intermediate shaft has inadvertently been placed 90° in advance of the correct position. The driving shaft speed is 200 RPM.

1. Follow the derivation of equation (4.22).
2. As shown in Figures 4.22a and b, the two forks at the end of the intermediate shaft should be in the same plane and the inclination angles at both ends should also be the same. Make the derivation given in equation (4.28) to show that under these conditions the output shaft speed is the same as that of the input shaft.
3. If the two end forks of the intermediate shaft are set at right angles by mistake, then

$$\frac{\omega_b}{\omega_a} = \frac{1}{\cos^2\delta\cos^2\alpha + \frac{\sin^2\alpha}{\cos^2\delta}}$$

   Therefore, the minimum speed of the output shaft occurs at $a = 90°$, i.e., $\omega_{b\,\min} = 200 \times \cos^2 20° = 176.6$ RPM.
4. The maximum speed of the output shaft occurs when $a = 0°$, i.e.,

$$\omega_{b\,\max} = \frac{200}{\cos^2 20°} = 226.5 \text{ RPM}$$

Solved Problem 4.12

Two shafts, the axes of which intersect but are inclined at 20° to each other are connected by a Hooke's joint. If the driving shaft has a uniform speed of 1000 RPM, find from first principles the variation in the speed of the driven shaft. The driven shaft carries a rotating mass of 15 kg having a radius of gyration 250 mm. Find the accelerating torque on the driven shaft for the position when the driven shaft has turned through 45° from the position in which its fork end is in the plane containing two shafts.

Follow the derivation of equation (4.21).

1. From equation (4.19), the angle of the driving shaft α can be obtained by substituting for the given angle β of the driven shaft and the inclination angle. $\tan\alpha = \tan 45° \cos 20°$, i.e., $\alpha = 43.22°$.
2. The speed of the driving shaft is $\omega_a = 2\pi(1000/60) = 104.7198$ rad/s.
3. From equation (4.21), the speed of the driven shaft is given by

$$\frac{\omega_b}{\omega_a} = \frac{\cos\delta}{1 - \sin^2\delta\cos^2\alpha}$$

Therefore

$$\omega_b = \frac{104.7198\cos 20°}{1 - \sin^2 20°\cos^2 43.22°} = \frac{104.7198 \times 0.9397}{1 - 0.117 \times 0.531} = 104.92 \text{ rad/s}$$

4. The angular acceleration of the driven shaft is given by equation (4.24) and upon substitution of appropriate values we get

$$\omega_b = -104.92^2\left[\frac{0.9397 \times 0.117 \times 0.9981}{(1 - 0.117 \times 0.531)^2}\right] = -13.68.1 \text{ rad/s}$$

Therefore the accelerating torque is $T = 15 \times 0.25^2 \times 1368.1 = 1282.6$ Nm.

Solved Problem 4.13

Show that for two shafts connected by a Hooke's joint, the ratio of angular velocities is given by

$$\frac{\omega_2}{\omega_1} = \frac{\cos\alpha}{1 - \sin^2\theta\sin^2\alpha}$$

where θ is the angle of rotation of shaft 1 from the position where the forked end is perpendicular to the plane containing the shaft axes and α is the angle of deviation of the drive.

Two shafts A and B are connected by a Hooke's joint, the angle of deviation being 20°. When the shafts are set in motion and allowed to rotate freely without friction, find the fluctuation of speed of shaft A as a percentage of its mean speed.

1. The equation for the ratio of angular velocities is derived, see equation (4.21). Note that α in equation (4.21) is equal to $90° - \theta$ and the shaft inclination angle there is denoted by δ. With the changed notation, you can derive the required relation.
2. The fluctuation in speed is given by equation (4.22) $\sin 20° \tan 20° = 0.1245$.

Solved Problem 4.14

For two shafts connected by a Hooke's joint, show that, if shaft 1 has a uniform angular velocity ω_1, the angular acceleration of shaft 2 is given by

$$\dot{\omega}_2 = -\omega_1^2\left[\frac{\cos\alpha\sin^2\alpha\sin 2\theta_1}{(1 - \sin^2\alpha\cos^2\theta_1)^2}\right]$$

where θ_1 is the angle of rotation of shaft 1 from the position where its forked end is in the plane containing the shaft and α is the angle of deviation of the drive. In a particular case, shaft 1 is driven at a constant speed of 500 RPM. Shaft 2 carries a mass moment of inertia 1 kg m^2 and is subjected to a constant resisting torque of 2750 Nm. Find the torque input to shaft 1 at the instant when $\theta_1 = 30°$. The inclination angle is 25°.

Follow the derivation of equation (4.24)

500 RPM = 52.36 rad/s

$$\dot{\omega}_2 = -52.36^2 \left[\frac{\cos 25° \sin^2 25° \sin 60°}{(1 - \sin^2 25° \cos^2 30°)^2} \right] = -512.41 \text{ rad/s}^2$$

Therefore the torque to accelerate the driving shaft is

$$T_{\text{inertia}} = 1 \times 512.41 = 512.41 \text{ Nm}$$

$$T_2 = 2750 - 512.41 = 2862.41 \text{ Nm}$$

$$T_1 = T_2 \frac{\omega_2}{\omega_1} = 2862.41 \frac{\cos\alpha}{1 - \sin^2\alpha\cos^2\theta_1} = 2862.41 \frac{\cos 25°}{1 - \sin^2 25° \cos^2 30°}$$

$$= 2995.4 \text{ Nm}$$

Solved Problem 4.15

Two parallel shafts are connected by an intermediate shaft with a Hooke's joint at each end. Show how the joints should be arranged to obtain a constant velocity ratio between the driving and driven shafts.

The intermediate shaft has a mass moment of inertia of 3 g m^2 and is inclined at 30° to the axes of the driving and driven shafts. If the driving shaft rotates uniformly at 2400 RPM with a steady input torque of 300 Nm, determine the maximum fluctuation of the output torque.

1. As shown in Figures 4.22a and b, the two forks at the end of the intermediate shaft should be in the same plane and the inclination angles at both ends should also be the same. Make the derivation given in equation (4.28) to show that under these conditions the output shaft speed is the same as that of the input shaft.
2. Maximum acceleration in the intermediate shaft occurs, see equation (4.25), when
$$\cos 2\alpha = \frac{2\sin^2 30°}{2 - \sin^2 30°} = 0.2857$$
i.e., when $\alpha = 36.7°$ and 216.7°.
3. Maximum acceleration from equation (4.24) is
$$\dot{\omega}_b = -\omega_a^2 \left[\frac{\cos\delta \sin^2\delta \sin 2\alpha}{(1 - \sin^2\delta\cos^2\alpha)^2} \right]$$
4. $\omega_a = 2\pi(2400/60) = 251.328$ rad/s.

5. $\dot{\omega}_b = -251.328^2 \left[\frac{0.866 \times 0.25 \times 0.9583}{(1 - 0.25 \times 0.6428)^2} \right] = -18605 \text{ rad/s}^2.$
6. For $\alpha = 143.3°$ and $323.3°$, $\dot{\omega}_b = 18605 \text{ rad/s}^2$.
7. Steady torque in the system is 300 Nm.
8. Maximum torque in the intermediate shaft due to fluctuation in acceleration, $T_{c\,max}$ is given by $0.003 \times 18605 = 55.815$ Nm.
9. Similarly, the minimum torque $T_{c\,min} = -55.815$ Nm.
10. The mean torque of the output shaft torque is same as the input shaft torque. Therefore, the output shaft torque will fluctuate between 355.815 and 244.185 Nm, with a maximum fluctuation 111.63 Nm.

Solved Problem 4.16

In a single Hooke's joint, the angle of divergence is 22.5° and a steady torque 250 Nm is applied to the driving shaft while it is rotating at 120 RPM. What must be the mass of a flywheel, whose radius of gyration is 250 mm, attached to the driven shaft so that the output torque does not vary by more than 25%?

1. Maximum acceleration in the intermediate shaft occurs, see equation (4.25), when
$$\cos 2\alpha = \frac{2 \sin^2 22.5°}{2 - \sin^2 22.5°} = 0.158$$
i.e., when $\alpha = 40.454°$ and $220.514°$.
2. Maximum acceleration from equation (4.24) is
$$\dot{\omega}_b = -\omega_a^2 \left[\frac{\cos \delta \sin^2 \sin 2\alpha}{(1 - \sin^2 \delta \cos^2 \alpha)^2} \right]$$
3. $\omega_a = 2\pi \frac{120}{60} = 12.5664 \text{ rad/s}.$
4. $\dot{\omega}_b = -12.5664^2 \left[\frac{0.9239 \times 0.1464 \times 0.9874}{(1 - 0.1464 \times 0.579)^2} \right] = -25.186 \text{ rad/s}^2.$
5. For $\alpha = 139.546°$ and $319.546°$, $\dot{\omega}_b = 25.186 \text{ rad/s}^2$.
6. From equation (4.21) the speed of the output shaft when maximum acceleration occurs is given by
$$\omega_b = \frac{\omega_a \cos 22.5°}{1 - \sin^2 22.5° \cos^2 40.45°} = 1.0095\omega_a$$
7. The torque transmitted at this speed is $250/1.0095 = 247$.
8. The output torque does not vary by more than 25%, therefore the lowest value of the output torque is three fourths of $250 = 187.5$ Nm. The accelerating torque is the transmitted torque minus the output torque, i.e., $247.65 - 187.5 = 60.15$ Nm.
9. Therefore, $60.15 = m \times 0.0625 \times 25.186$ or $m = 38.21$ kg.

4.7 Additional Problems

1. Derive the condition for exact straight line motion. Show that the Paucellier mechanism satisfies this condition.
2. Show that the Hart linkage generates an exact straight line.
3. State the conditions for a Scott–Russel mechanism to generate an exact straight line. Describe how such a mechanism can be modified by replacing the slider with a rocker and generate an approximate straight line. Modify this linkage further so that the approximate straight line generated is offset from the crank rotation center. Obtain the best point to describe an approximate straight line for such a linkage by using the concept of instantaneous center.
4. A modified Scott–Russel mechanism has its crank, connecting rod, rocker and ground links of lengths 1.65, 2, 2 and 4 units respectively. Draw the linkage when the crank makes an angle of 60° from the base link. Find the best point on the coupler that will describe an offset approximate straight line. Draw the approximate straight line expected at this position of the linkage. Also find the coupler curve equation for this point.
5. In a Watt's linkage, the two rocking levers are in the ratio of 4 to 3 units. Show how you can obtain approximate straight line motion for a suitable point on the coupling rod. The coupler rod can be around 1.5 to 2 units in length.
6. State the conditions for generating an approximate straight line of a point on the coupler rod of a Tchebicheff linkage. Using suitable proportions draw the linkage and show the points through which the straight line will pass. Derive the coupler curve equation of the point describing the approximate straight line.
7. Design a Davis steering gear linkage for an automobile with a wheel base 2.5 m. The distance between the pivots of the front axle may be taken to be around 1 m. Derive any relations used in the calculations.
8. An Ackermann steering gear mechanism is used for an automobile with a track of 1.5 m and a wheel base of 2.75 m. Design the four-bar linkage to give the best possible steering. You may take the distance between the front stub axle pivots to be around 1.25 m. Make a plot of $\cot\phi - \cot\theta$ as a function of θ and show where the correct steering occurs.
9. Two horizontal shafts A and B whose axes intersect and are inclined at an angle δ are connected by a Hooke's joint. If A is the driving shaft and rotates at a uniform angular speed ω rad/s, find an expression for the angular speed of the shaft B when one of the cross arms A is turned through an angle θ from its vertical position. Hence, determine for the case when $\delta = 30°$ the maximum and minimum velocity ratios and the ratio of the fluctuation of the speed of B to the mean speed.
10. Two horizontal shafts are connected by a Hooke's joint. The angle between the shafts is 160°. The driving shaft rotates uniformly at 150 RPM. The driven shaft carries a flywheel weighing 10 kg having a radius of gyration 10 cm. Find the torque required on the driving shaft to overcome the inertia of the flywheel, when the fork end of the driving shaft has rotated 30° from the horizontal plane.

Chapter 5
Cams

Cam

Component with a curved profile or surface whereby it imparts a displacement either by point or line contact with a cam follower.

Cam Follower

Component that receives motion directly from a cam.

5.1 Types of Cams and Followers

Disk [Plate or Radial] Cam

Disk that rotates about an axis perpendicular to its plane and drives a follower through contact with its profile, see Figure 5.1.

Cylindrical [Barrel] Cam

Rotating cylinder with a curved groove in its surface or a curved rib on its surface whereby contact is made with a follower, see Figure 5.2.

Translation Cam

Cam with a translatory motion having a profile on one side whereby contact is made with a follower, see Figure 5.3.

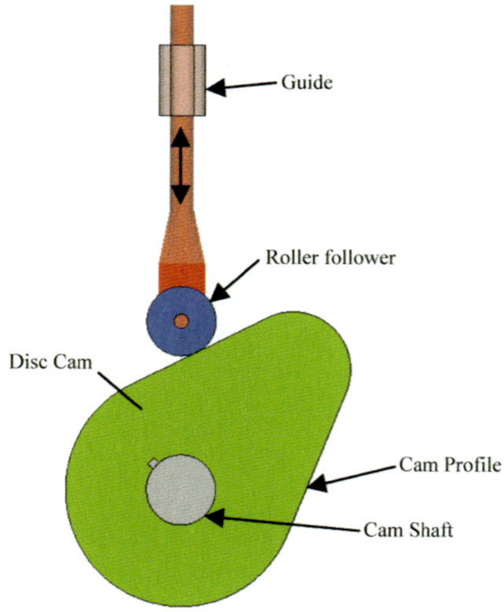

Fig. 5.1 Disk cam with translating roller follower

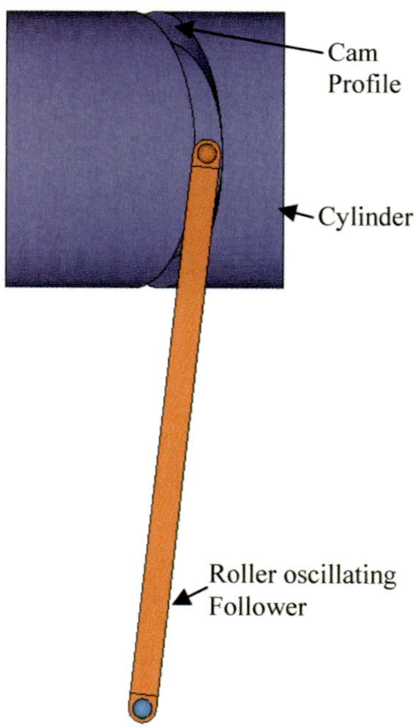

Fig. 5.2 Cylindrical cam

Face Cam

Rotating cam that makes contact with a follower by means of a groove in or a rib on a plane surface that is perpendicular to the axis of cam, see Figure 5.4.

Spherical Cam

Rotating hollow sphere with a groove in or a rib on its inner surface to make contact with a follower.

Yoke Cam

Constant-breadth radial cam designed to mesh with a yoke follower.

Translatory Follower

Follower receiving translatory motion that passes through the center of rotation of a cam; see Figure 5.5a.

Knife Edge Follower

Follower with a knife edge point that makes contact with and receives motion from a cam; see Figure 5.6a.

Roller Follower

A roller on the follower makes contact with and receives motion from a cam, see Figure 5.1.

Flat Follower

A flat surface on the follower makes contact with and receives motion from a cam; see Figure 5.5a.

Spherical Faced Follower

A spherical faced surface on the follower makes contact with and receives motion from a cam; see Figure 5.5b.

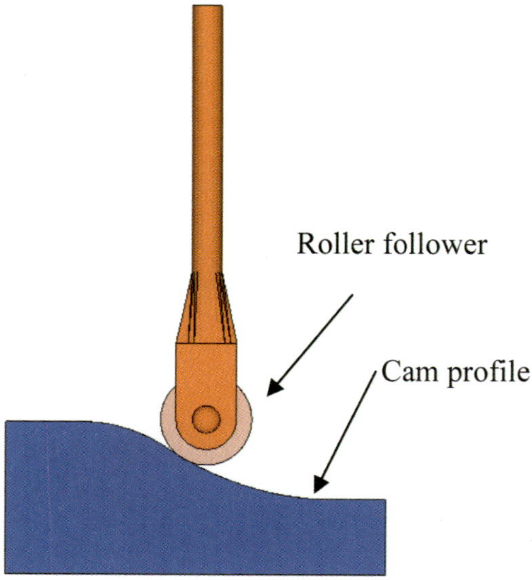

Fig. 5.3 Translation cam

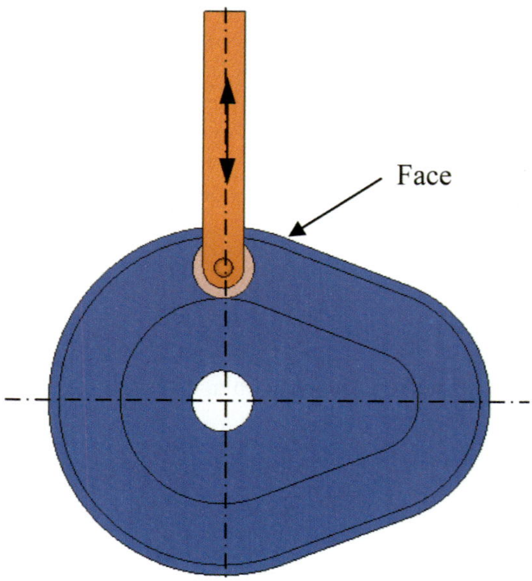

Fig. 5.4 Face cam

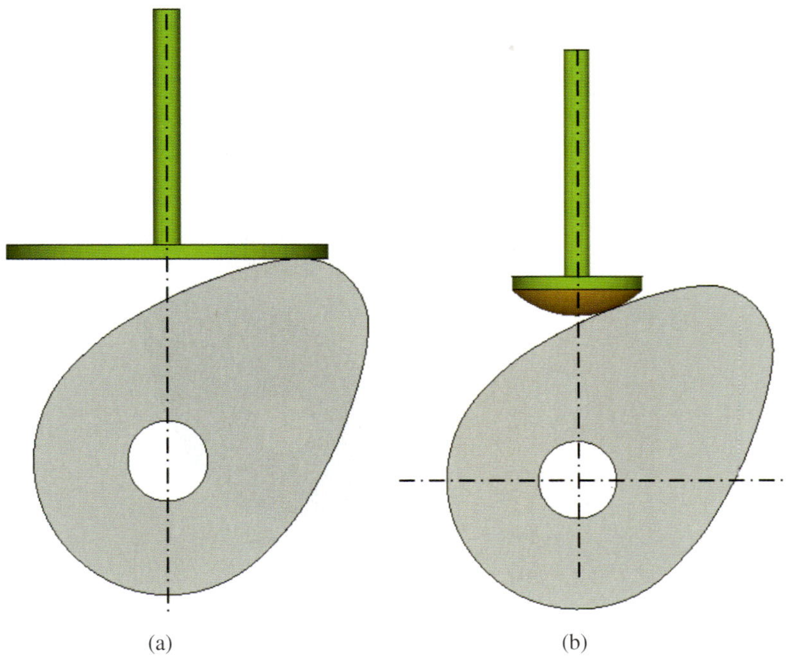

Fig. 5.5 (a) Translatory follower through the center of a cam. (b) Spherical faced translatory follower

Offset Follower

A follower receiving translatory motion that is offset from the center of rotation of a cam, see Figures 5.6b and c.

Oscillating Follower

A follower receives an oscillatory motion from a cam, see Figure 5.7.

The cam followers can be classified depending on the type of element that receives different types of motion, e.g., translatory roller follower, translatory offset flat follower, oscillating roller follower, etc.

5.2 Displacement Diagrams

Cam-follower systems are designed to achieve a desired oscillatory motion as in the case of internal combustion engine valves, machine tool drives, printing machinery, computer drives, etc. Appropriate displacement diagrams are to be chosen for this

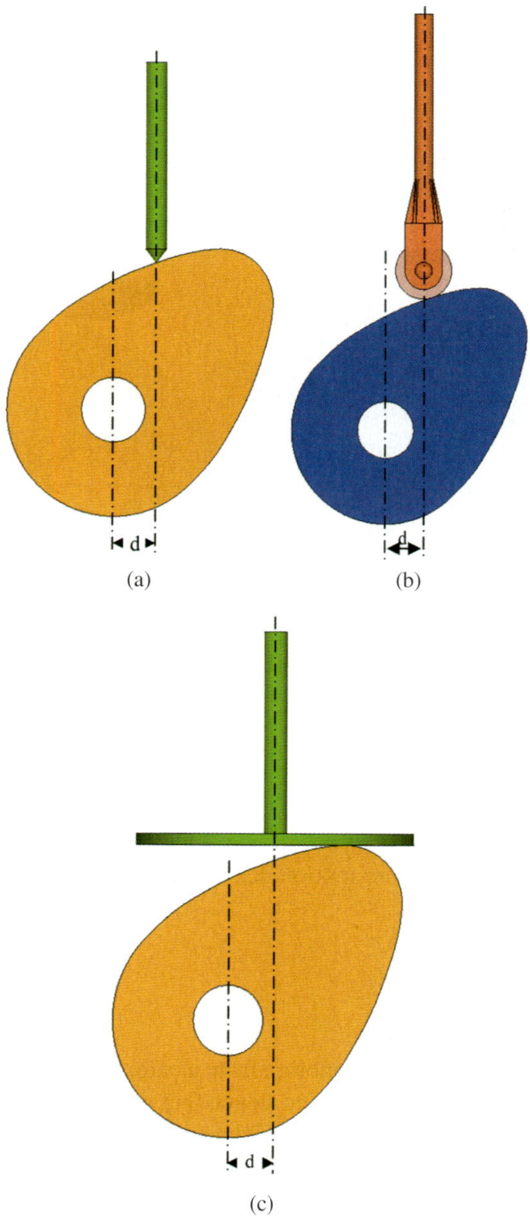

Fig. 5.6 (a) Translating knife edge offset follower. (b) Translating roller offset follower. (c) Translating flat offset follower

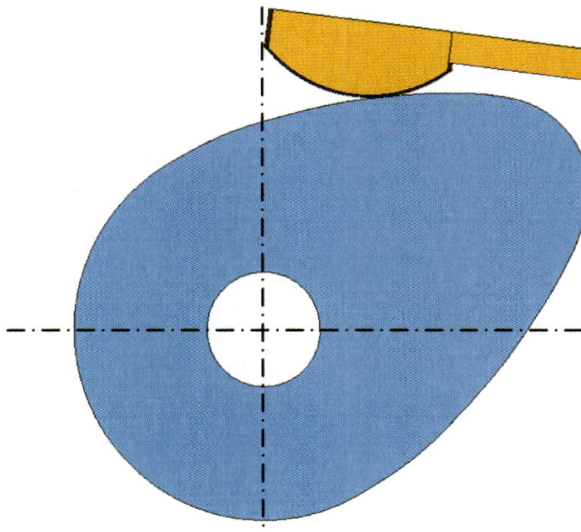

Fig. 5.7 Oscillating follower

purpose before designing the cam surface. The cam is assumed to rotate at a constant speed and the follower in each revolution raises, dwells, and returns or falls to its original position and dwells again through specified angles of rotation of the cam before rising again in the next cycle. Various types of displacement diagrams for rise motion are described below.

Uniform [Constant Velocity] Motion

The total cam angle β and the lift L are divided into equal numbers of parts and the follower displacement at any cam angle θ, say 2 is given by the point $2'$, see Figure 5.8a. Analytically, the displacement y can be represented as

$$y = c\theta \tag{5.1}$$

The constant c in the above is determined from the boundary conditions

$$\begin{aligned} y &= 0 \quad \text{at} \quad \theta = 0 \\ y &= L \quad \text{at} \quad \theta = \beta \end{aligned} \tag{5.2}$$

Therefore

$$y = \frac{L}{\beta}\theta \tag{5.3}$$

The velocity and acceleration are obtained by taking the derivative of y with time and using $(\omega = d\theta/dt)$

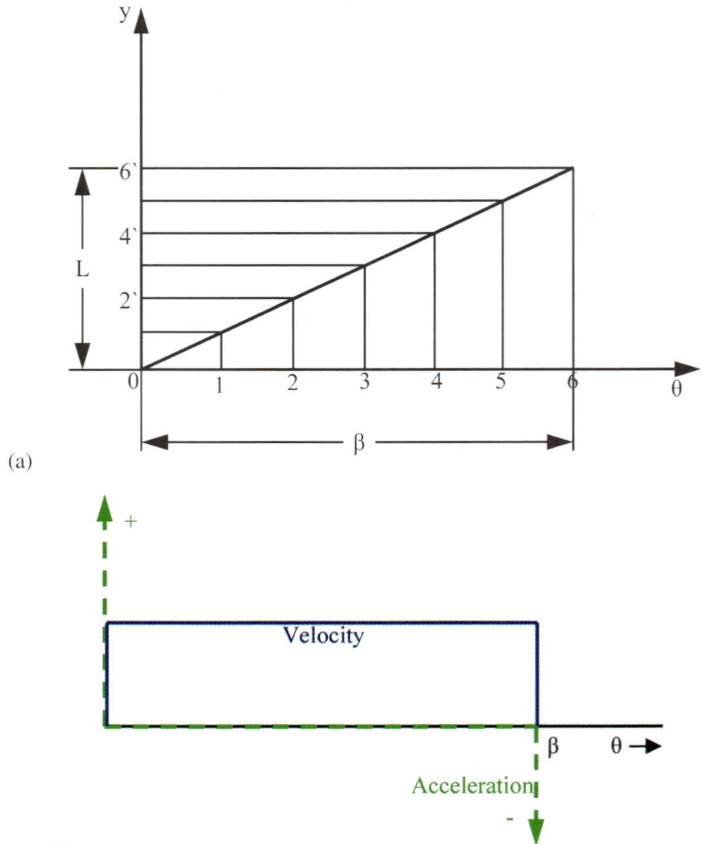

Fig. 5.8 (a) Uniform (constant velocity) motion. (b) Velocity and acceleration for uniform motion

$$v = \frac{dy}{dt} = \frac{d}{dt}\left(\frac{L}{\beta}\theta\right) = \frac{L}{\beta}\omega \tag{5.4}$$

$$a = \frac{d^2y}{dt^2} = 0 \tag{5.5}$$

Note that the velocity jumps from 0 to a constant value suddenly at time $t = 0$ and drops to zero again when $\theta = \beta$. Therefore

$$a|_{\theta=0} = +\infty \quad \text{and} \quad a|_{\theta=\beta} = -\infty \tag{5.6}$$

The velocity and accelerations are shown in Figure 5.8b. This motion is denoted by CV-1 for rise and CV-2 for fall motion.

We will learn later that the inertia force is a product of mass and acceleration and if acceleration reaches infinity these forces become infinity or very large; therefore

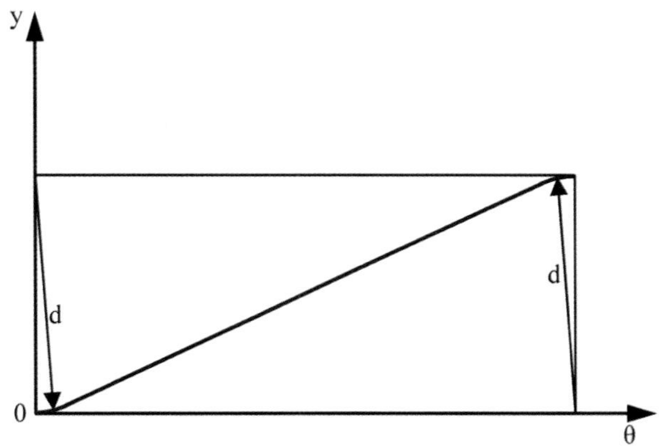

Fig. 5.9 Modified uniform motion

CV-1 and CV-2, though they are simple cannot be used even for fairly marginally high angular velocities. They are limited to very slow speeds.

Uniform Motion Modified by a Circular Arc

Because of infinite inertias at the beginning and completion of rise motion in Figure 5.8, the uniform motion is generally rounded off by a circular arc as shown in Figure 5.9. In early days such small modifications were used to retain simplicity and ease in manufacturing.

Parabolic [Constant Acceleration] Motion

Graphical construction of this motion can be made as follows.

Choose any convenient length to represent the cam rotation angle β and draw a vertical line to represent total lift L as shown in Figure 5.10a.

1. The cam angle is divided into equal parts, say six as shown in the figure. (This number is chosen for the purpose of illustration, otherwise technicians can do a better job in manufacturing with a larger number of divisions; alternatively numerically controlled machine tools can, in the present day, do a far more accurate job.)
2. Draw a convenient line and divide it in the ratio 1:3:5:5:3:1 (or 1:3:5:7:7:5:3:1 if the cam angle is divided into eight parts, etc.).
3. The total lift is now divided in the same ratios by drawing parallel lines as shown.
4. Mark the corresponding points, $1'$-1, $2'$-2, …, $6'$-6 and join them by a smooth line.

Analytically, the parabolic motion can be represented by

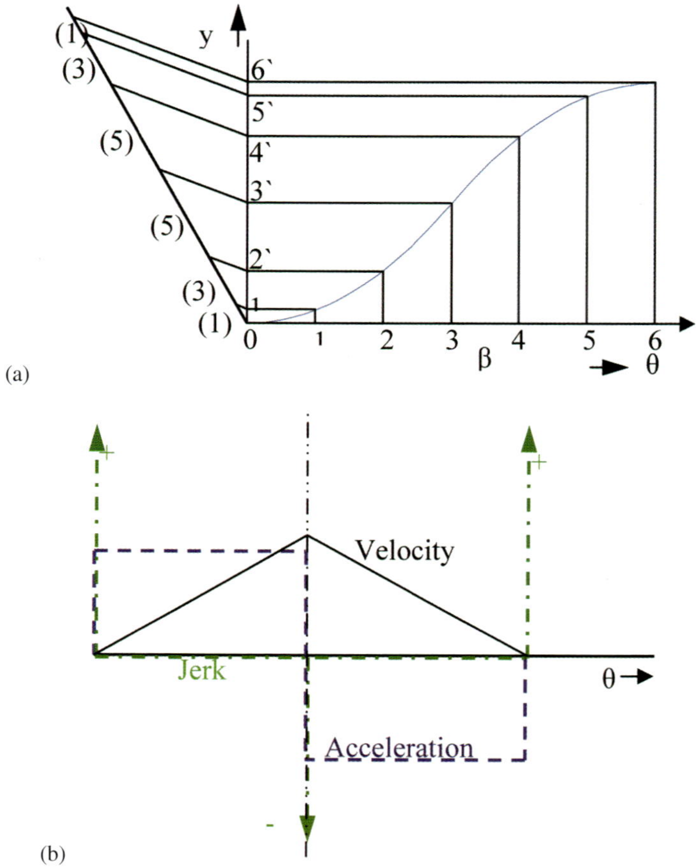

Fig. 5.10 (a) Parabolic motion. (b) Velocity, acceleration and jerk of parabolic motion

$$y = c\theta^2 \tag{5.7}$$

upto the inflection point 3 in Figure 5.10. This is an important point, where the maximum steepness of the cam is defined. The boundary conditions are

$$y = 0 \text{at} \theta = 0$$

$$y = \frac{1}{2}L \quad \text{at} \quad \theta = \frac{1}{2}\beta \tag{5.8}$$

Note that the number of boundary conditions is limited depending on the unknowns in equation (5.7). We take whatever velocities and accelerations result in such a motion and cannot control them with simple motions such as CV-1, CV-2 or parabolic motion. Therefore

$$y = 2L\left(\frac{\theta}{\beta}\right)^2 \tag{5.9}$$

The velocity and acceleration for this region are

$$v = \frac{4L\omega}{\beta^2}\theta \tag{5.10}$$

$$a = \frac{4L\omega^2}{\beta^2} \tag{5.11}$$

Since the velocity increases uniformly with constant acceleration, parabolic motion is often referred to as Constant Acceleration Motion. The maximum velocity occurs at the midpoint of the cam angle when $\theta = 1/2\beta$, which is

$$v\big|_{\theta=\frac{1}{2}\beta} = \frac{2L\omega}{\beta} \tag{5.12}$$

For the second half of the motion we write the full second-degree equation for a parabola

$$y = c_1 + c_2\theta + c_3\theta^2 \tag{5.13}$$

The boundary conditions are

$$\begin{aligned} y &= \frac{1}{2}L \quad \text{at} \quad \theta = \frac{1}{2}\beta, \quad \dot{y} = \frac{2L\omega}{\beta} \\ y &= L \quad \text{at} \quad \theta = \beta \end{aligned} \tag{5.14}$$

Therefore, the constants in equation (5.13) are

$$\begin{aligned} c_1 &= -L \\ c_2 &= \frac{4L}{\beta} \\ c_3 &= -\frac{2L}{\beta^2} \end{aligned} \tag{5.15}$$

Hence we can write the lift, velocity and acceleration as functions of rotation angle θ:

$$y = L\left[1 - 2\left(1 - \frac{\theta}{\beta}\right)^2\right] \tag{5.16}$$

$$v = \frac{4L\omega}{\beta}\left(1 - \frac{\theta}{\beta}\right) \tag{5.17}$$

$$a = -\frac{4L\omega^2}{\beta^2} \tag{5.18}$$

We note that the acceleration is constant in the first and second halves of the motion. In the first half, it rises to a positive acceleration $4L\omega^2/\beta^2$ from zero and at its inflection point it changes to deceleration $-4L\omega^2/\beta^2$ before returning to zero. The rate of change of acceleration is defined as jerk J and is given for parabolic motion by

$$\theta = 0, \quad J = +\infty$$
$$\theta = \frac{1}{2}\beta, \quad J = -\infty$$
$$\theta = \beta, \quad J = +\infty \tag{5.19}$$

Figure 5.10b gives the above relations. Parabolic motion for rise is denoted as PB-1 and for fall it is denoted by PB-2. Though parabolic motion gives finite acceleration, therefore, finite inertia forces, there are three infinite jerks in each revolution which cause undue vibrations for high speed cams.

Trapezoidal Acceleration

The square wave's discontinuities of acceleration in Figure 5.10b can be removed by linearly increasing the acceleration in a cam angle $\beta/8$ at the start and finish of the rise motion and decreasing the acceleration linearly through an angle $\beta/4$ at the middle. Such a motion is denoted TP-1 for rise and TP-2 for fall motions. This motion removes infinite jerks and makes them finite three times in each rise or fall motion.

Modified Trapezoidal Acceleration

Here, the linear variation of acceleration or deceleration used in trapezoidal acceleration motion is replaced by split sine waves at the beginning, middle and end portions of the constant acceleration motion. The displacement, velocity, acceleration and jerk for this motion in different regions of the cam angle are given below.

$$\text{Region } 0 \le \theta < \frac{1}{8}\beta$$
$$y = L\left[0.38898448\frac{\theta}{\beta} - 0.0309544\sin\frac{4\pi\theta}{\beta}\right]$$
$$v = 0.38898448\frac{L\omega}{\beta}\left[1 - \cos\frac{4\pi\theta}{\beta}\right]$$
$$a = 4.888124\frac{L\omega^2}{\beta^2}\sin\frac{4\pi\theta}{\beta}$$
$$J = 61.425769\frac{L\omega^3}{\beta^3}\cos\frac{4\pi\theta}{\beta} \tag{5.20}$$

Region $\frac{1}{8}\beta \leq \theta < \frac{3}{8}\beta$

$$y = L\left[2.44406184\left(\frac{\theta}{\beta}\right)^2 - 0.22203097\frac{\theta}{\beta} + 0.00723407\right]$$

$$v = \frac{L\omega}{\beta}\left[4.888124\frac{\theta}{\beta} - 0.22203097\right]$$

$$a = 4.888124\frac{L\omega^2}{\beta^2}$$

$$J = 0 \tag{5.21}$$

Region $\frac{3}{8}\beta \leq \theta < \frac{5}{8}\beta$

$$y = L\left[1.6110154\frac{\theta}{\beta} - 0.0309544\sin\left(\frac{4\pi\theta}{\beta} - \pi\right) - 0.3055077\right]$$

$$v = \frac{L\omega}{\beta}\left[1.6110154 - 0.38898448\cos\left(\frac{4\pi\theta}{\beta} - \pi\right)\right]$$

$$a = 4.888124\frac{L\omega^2}{\beta^2}\sin\left(\frac{4\pi\theta}{\beta} - \pi\right)$$

$$J = 61.425769\frac{L\omega^3}{\beta^3}\cos\left(\frac{4\pi\theta}{\beta} - \pi\right) \tag{5.22}$$

Region $\frac{5}{8}\beta \leq \theta < \frac{7}{8}\beta$

$$y = L\left[-2.44406184\left(\frac{\theta}{\beta}\right)^2 + 4.6660917\frac{\theta}{\beta} - 1.2292648\right]$$

$$v = \frac{L\omega}{\beta}\left[-4.888124\frac{\theta}{\beta} + 4.6660917\right]$$

$$a = -4.888124\frac{L\omega^2}{\beta^2}$$

$$J = 0 \tag{5.23}$$

Region $\frac{7}{8}\beta \le \theta < \beta$

$$y = L\left[0.38898448\frac{\theta}{\beta} + 0.0309544\sin\left(\frac{4\pi\theta}{\beta} - 3\pi\right) + 0.6110154\right]$$

$$v = 0.38898448\frac{L\omega}{\beta}\left[1 + \cos\left(\frac{4\pi\theta}{\beta} - 3\pi\right)\right]$$

$$a = -4.888124\frac{L\omega^2}{\beta^2}\sin\left(\frac{4\pi\theta}{\beta} - 3\pi\right)$$

$$J = -61.425769\frac{L\omega^3}{\beta^3}\cos\left(\frac{4\pi\theta}{\beta} - 3\pi\right) \tag{5.24}$$

Simple Harmonic [Cosine Acceleration] Motion

The displacement curve is shown in Figure 5.11a. The following steps may be followed to construct this curve:

1. Draw a semi circle with lift L as diameter as shown in Figure 5.11a.
2. Divide the circle into the same number of parts as the cam angle β is divided.
3. Draw horizontal lines from the peripheral divisions of the circle to obtain $1'$, $2'$, $\ldots$, $6'$.
4. Locate the points of the curve for different positions of cam angle 1, 2, ..., 6 and draw a smooth curve through these points as shown.

We call this a full harmonic and denote it as H-5. The equations for displacement, velocity, acceleration and jerk relations for such a harmonic are

$$y = \frac{L}{2}\left(1 - \cos\frac{\pi\theta}{\beta}\right)$$

$$v = \frac{\pi L\omega}{2\beta}\sin\frac{\pi\theta}{\beta}$$

$$a = \frac{L}{2}\left(\frac{\pi\omega}{\beta}\right)^2\cos\frac{\pi\theta}{\beta}$$

$$J = -\frac{L}{2}\left(\frac{\pi\omega}{\beta}\right)^3\sin\frac{\pi\theta}{\beta} \quad \text{for} \quad 0 < \theta < \beta$$

$$J = +\infty \quad \text{for} \quad \theta = 0 \text{ and } \beta \tag{5.25}$$

Since the acceleration is the cosine, the harmonic motion is also called "cosine acceleration curve". Even though there is a finite acceleration, there are still two jerks in every revolution. The corresponding motion for fall is denoted by H-6.

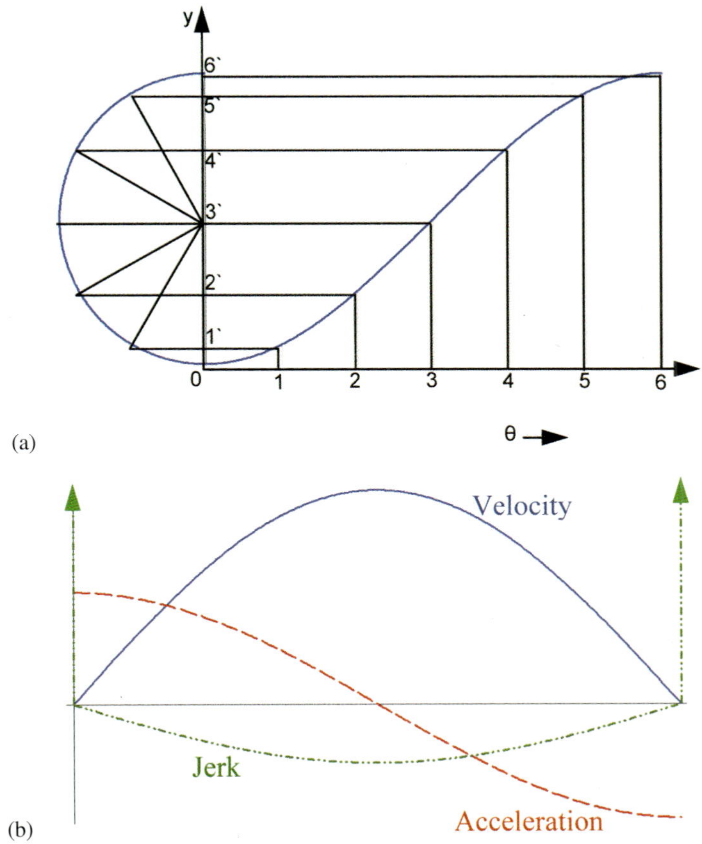

Fig. 5.11 (a) Simple harmonic (cosine acceleration or H-5) motion. (b) Velocity, acceleration and jerk of simple harmonic motion

Cycloidal [Sine Acceleration] Motion

The construction procedure for this lift curve is as follows:

1. Construct a circle of radius $L/2\pi$ with its center at full lift position as shown in Figure 5.12a.
2. Divide the circle into the same number of parts as the cam angle has been divided, in this case, six as shown in the figure.
3. Name the points thus obtained as $1'$, $2'$, ..., $6'$ in a clockwise manner beginning from position 0 as shown in the figure.
4. Project these points onto the vertical line.
5. Join point 0 on the cam angle axis with the projected point $0'$, which is the diagonal line in Figure 5.12a.

(a)

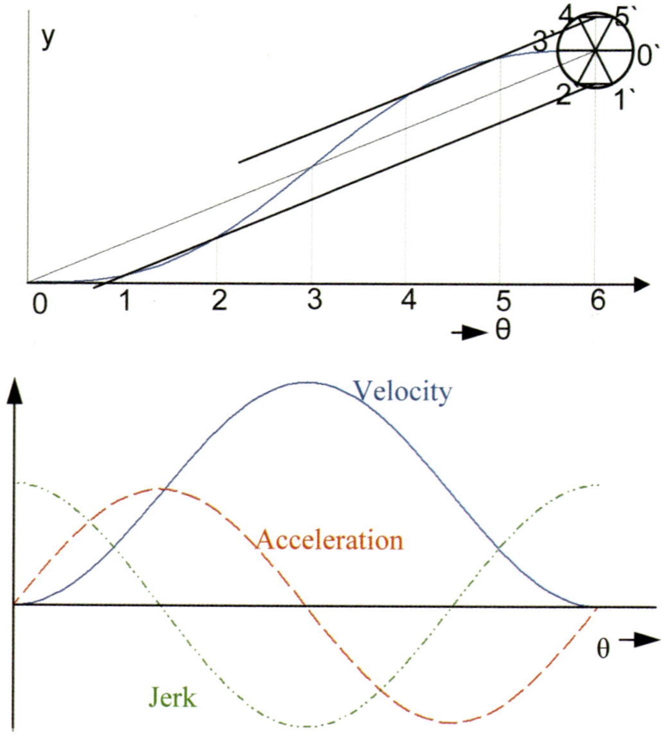

(b)

Fig. 5.12 (a) Cycloidal (sine acceleration or C-5) motion. (b) Velocity, acceleration and jerk of cycloidal motion

6. Draw parallel lines from the projected points of $1', 2', \ldots, 6'$ to the vertical line to intersect the vertical lines at the corresponding positions $1, 2, \ldots, 6$ from the cam angle axis.
7. Draw a smooth curve through these intersection points to give the cycloidal lift curve.

This motion is a full cycloid, denoted by C-5. The equation for this motion along with its velocity, acceleration and jerk are given below:

$$y = L\left(\frac{\theta}{\beta} - \frac{1}{2\pi}\sin\frac{2\pi\theta}{\beta}\right)$$

$$v = \frac{L\omega}{\beta}\left(1 - \cos\frac{2\pi\theta}{\beta}\right)$$

$$a = 2L\pi\left(\frac{\omega}{\beta}\right)^2 \sin\frac{2\pi\theta}{\beta}$$

$$J = 4L\pi^2\left(\frac{\omega}{\beta}\right)^3 \cos\frac{2\pi\theta}{\beta} \tag{5.26}$$

This is also called "sine acceleration motion". (One can begin with acceleration expression $a = C \sin 2\pi(\theta/\beta)$, integrate and use the appropriate boundary conditions to derive the rest of the relations in the above equation). In this case, the jerk is a finite quantity throughout the cam angle and therefore is more suitable for high speed applications to keep the vibrations low. The corresponding motion for the fall is denoted by C-6.

Modified Sine Acceleration Motion

Here two sine waves for acceleration are used; one of a period $\beta/2$ and the other of period $3\beta/2$. The first and fourth quarters of the first sine curve are fitted at the first and last $\beta/8$ of the cam angle. The second and third quarters of the second sine curve is placed in the slot $\beta/8$ to $7\beta/8$ of the cam angle. The equations for the acceleration of rise motion are given by

Region $0 \le \theta < \frac{1}{8}\beta$

$$y = L\left[0.43990085\frac{\theta}{\beta} - 0.0350062\sin\frac{4\pi\theta}{\beta}\right]$$

$$v = 0.43990085\frac{L\omega}{\beta}\left[1 - \cos\frac{4\pi\theta}{\beta}\right]$$

$$a = 5.5279571\frac{L\omega^2}{\beta^2}\sin\frac{4\pi\theta}{\beta}$$

$$J = 69.4663577\frac{L\omega^3}{\beta^3}\cos\frac{4\pi\theta}{\beta} \tag{5.27}$$

Region $\frac{1}{8}\beta \le \theta < \frac{7}{8}\beta$

$$y = L\left[0.28004957 + 0.43990085\frac{\theta}{\beta} - 0.31505577\cos\left(\frac{4\pi}{3}\frac{\theta}{\beta} - \frac{\pi}{6}\right)\right]$$

$$v = 0.43990085\frac{L\omega}{\beta}\left[1 + 3\sin\left(\frac{4\pi}{3}\frac{\theta}{\beta} - \frac{\pi}{6}\right)\right]$$

$$a = 5.5279571\frac{L\omega^2}{\beta^2}\cos\left(\frac{4\pi}{3}\frac{\theta}{\beta} - \frac{\pi}{6}\right)$$

$$J = -23.1533\frac{L\omega^3}{\beta^3}\sin\left(\frac{4\pi}{3}\frac{\theta}{\beta} - \frac{\pi}{6}\right) \tag{5.28}$$

Region $\frac{7}{8}\beta \leq \theta \leq \beta$

$$y = L\left[0.56009915 + 0.43990085\frac{\theta}{\beta} + 0.0350062\sin\frac{4\pi\theta}{\beta}\right]$$

$$v = 0.43990085\frac{L\omega}{\beta}\left[1 - \cos\frac{4\pi\theta}{\beta}\right]$$

$$a = 5.5279571\frac{L\omega^2}{\beta^2}\sin\frac{4\pi\theta}{\beta}$$

$$J = 69.4663577\frac{L\omega^3}{\beta^3}\cos\frac{4\pi\theta}{\beta} \tag{5.29}$$

Polynomial Motion

The above displacement curves are chosen from known mathematical functions. We have seen that some of them have infinite accelerations or jerks which cause severe stresses and vibration problems. Moreover, we may like to have a specified function for the displacement curve and it may produce sudden discontinuities in motion. Therefore, polynomial functions of a desired degree can be generated to suit the specific demands in a design problem. The simplest case of a polynomial function is

$$y = C_0 + C_1\theta \tag{5.30}$$

Since only two constants are chosen, we choose two boundary conditions to be satisfied. Let

$$y = 0 \quad \text{at} \quad \theta = 0$$

$$y = L \quad \text{at} \quad \theta = \beta \tag{5.31}$$

Therefore

$$y = \frac{L}{\beta}\theta \tag{5.32}$$

Next let us consider a third-degree polynomial with the boundary conditions (choosing four)

$$
\begin{aligned}
&y = C_0 + C_1\theta + C_2\theta^2 + C_3\theta^3 \\
&y = 0,\ \dot{y} = 0 \quad \text{at} \quad \theta = 0 \\
&y = L,\ \dot{y} = 0 \quad \text{at} \quad \theta = \beta
\end{aligned} \tag{5.33}
$$

The velocity of the follower for the motion in (5.33) is obtained by differentiating the same with respect to time t:

$$
\dot{y} = C_1\omega + 2C_2\omega\theta + 3C_3\omega\theta^2 \tag{5.34}
$$

The constants in equation (5.33) can now be obtained, which are

$$
\begin{aligned}
C_0 &= 0 \\
C_1 &= 0 \\
C_2 &= \frac{3L}{\beta^2} \\
C_3 &= -\frac{2L}{\beta^3}
\end{aligned} \tag{5.35}
$$

Therefore the displacement, velocity, acceleration and jerk for this motion are

$$
\begin{aligned}
y &= \frac{3L}{\beta^2}\theta^2 - \frac{2L}{\beta^3}\theta^3 \\
v &= \frac{6L}{\beta^2}\omega\theta - \frac{6L}{\beta^3}\omega\theta^2 \\
a &= \frac{6L}{\beta^2}\omega^2 - \frac{12L}{\beta^3}\omega^2\theta \\
J &= -\frac{12L}{\beta^3}\omega^3 \quad \text{for} \quad 0 < \theta < \beta \\
&= +\infty \quad \text{at} \quad \theta = 0 \\
&= -\infty \quad \text{at} \quad \theta = \beta
\end{aligned} \tag{5.36}
$$

Since the constants C_2 and C_3 are only present in the displacement curve, this cam is also called a 2-3 cam.

A 3-4-5 polynomial cam can be similarly derived from a fifth-degree polynomial with the boundary conditions given below:

$$
y = C_0 + C_1\theta + C_2\theta^2 + C_3\theta^3 + C_4\theta^4 + C_5\theta^5 \tag{5.37}
$$

$$y = 0, \quad \dot{y} = 0 \quad \text{and} \quad \ddot{y} = 0 \quad \text{at} \quad \theta = 0$$

$$y = L, \quad \dot{y} = 0 \quad \text{and} \quad \ddot{y} = 0 \quad \text{at} \quad \theta = \beta \tag{5.38}$$

The constants C can be obtained as

$$\begin{aligned} C_0 &= 0 \\ C_1 &= 0 \\ C_2 &= 0 \\ C_3 &= -\frac{10L}{\beta^3} \\ C_4 &= -\frac{15L}{\beta^4} \\ C_5 &= \frac{6L}{\beta^5} \end{aligned} \tag{5.39}$$

Finally,

$$\begin{aligned} y &= \frac{10L}{\beta^3}\theta^3 - \frac{15L}{\beta^4}\theta^4 + \frac{6L}{\beta^5}\theta^5 \\ v &= \frac{30L}{\beta^3}\omega\theta^2 - \frac{60L}{\beta^4}\omega\theta^3 + \frac{30L}{\beta^5}\omega\theta^4 \\ a &= \frac{60L}{\beta^3}\omega^2\theta - \frac{180L}{\beta^4}\omega^2\theta^2 + \frac{120L}{\beta^5}\omega^2\theta^3 \\ J &= \frac{60L}{\beta^3}\omega^3 - \frac{360L}{\beta^4}\omega^3\theta + \frac{360L}{\beta^5}\omega^3\theta^2 \end{aligned} \tag{5.40}$$

The advantage with the above type of cam is that it has finite and continuous accelerations and jerks throughout the cam angle for rise motion.

Combinations of Displacement Curves

A general procedure to choose displacement curves for rise and fall motions is to combine suitable motions from the curves discussed so far and other harmonic, cycloid and polynomial curves given in Figures 5.13, 5.14 and 5.15. H-1 to H-4 are half harmonics, the first two of them for rise motion and the next two for fall motion respectively. H-5 is a full harmonic discussed earlier and H-6 is for fall motion. Note that these curves have different end conditions for velocity and acceleration. Similarly Figure 5.14 gives C-1 to C-4 half Cycloidal motions and C-5 and C-6 full Cycloid motions. P-1 and P-2 are eighth-power polynomials for rise and fall motions respectively as given in Figure 5.15.

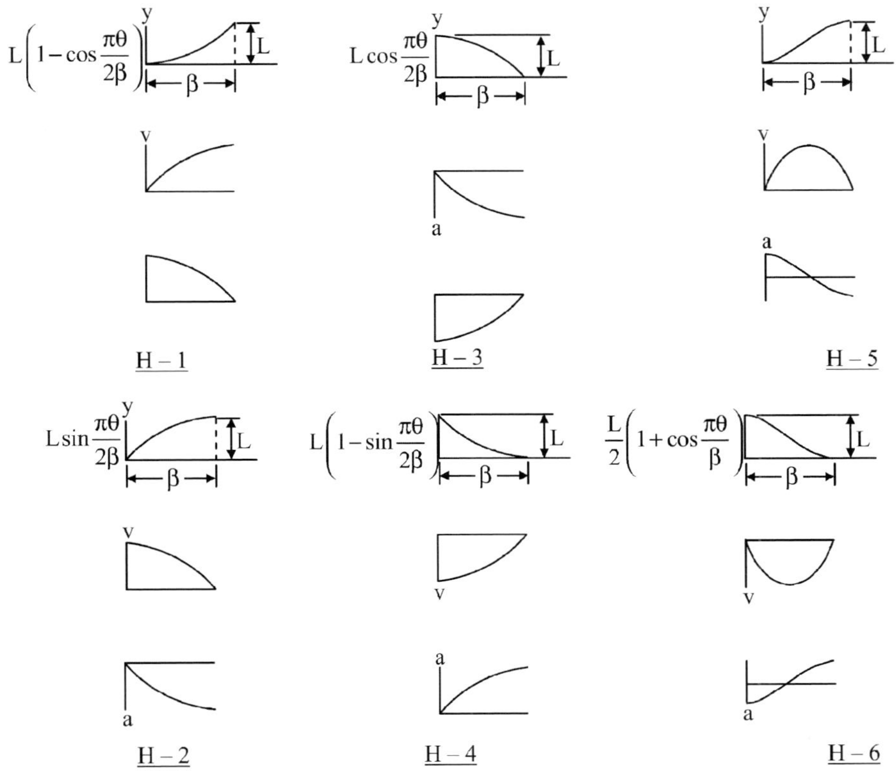

Fig. 5.13 Harmonic motion characteristics

As an example consider a follower that is required to dwell, rise with acceleration, rise with uniform velocity, further rise with deceleration and then dwell again according to the following requirements:

Point 1	Point 2	Point 3	Point 4
$\theta = 0$	$\theta = \theta_2$	$\theta = \theta_3$	$\theta = \theta_4$
$y = 0$	$y = L_1$	$y = L_1 + L_2$	$y = L_1 + L_2 + L_3$
$v = 0$	$v = v_1$	$v = v_1$	$v = 0$
$a = 0$	$a = 0$	$a = 0$	$a = 0$

Figure 5.16 gives a proposed solution. From point 1 to 2, C-1 half cycloid is chosen to provide zero acceleration at the beginning and completion, while connecting with uniform velocity of the next motion from point 2 to 3. From 3 to 4, half cycloid C-2 is used to couple the zero acceleration and constant velocity at the end of point 3 to bring the follower to dwell position. In the displacement, velocity and acceleration curves make sure there are no discontinuities.

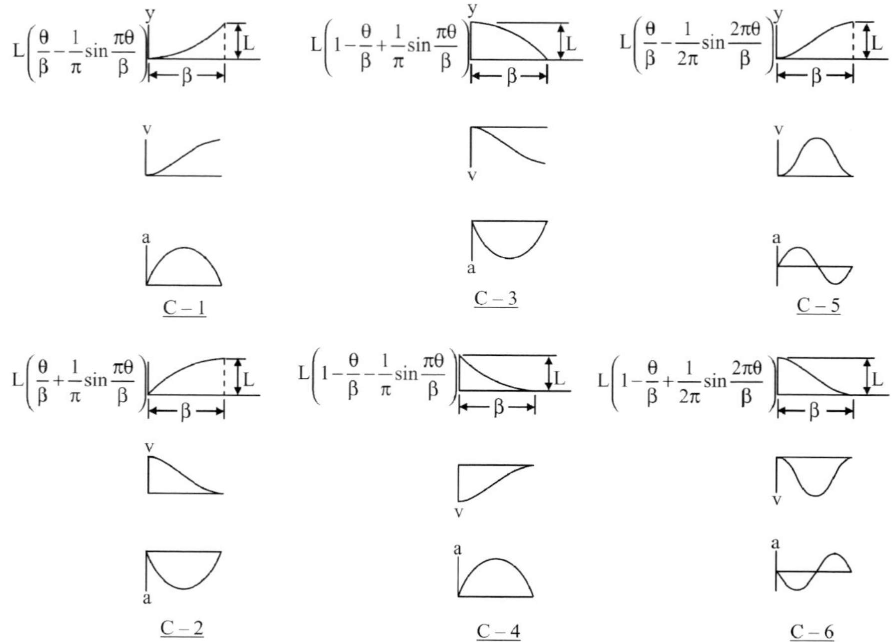

Fig. 5.14 Cycloidal motion characteristics

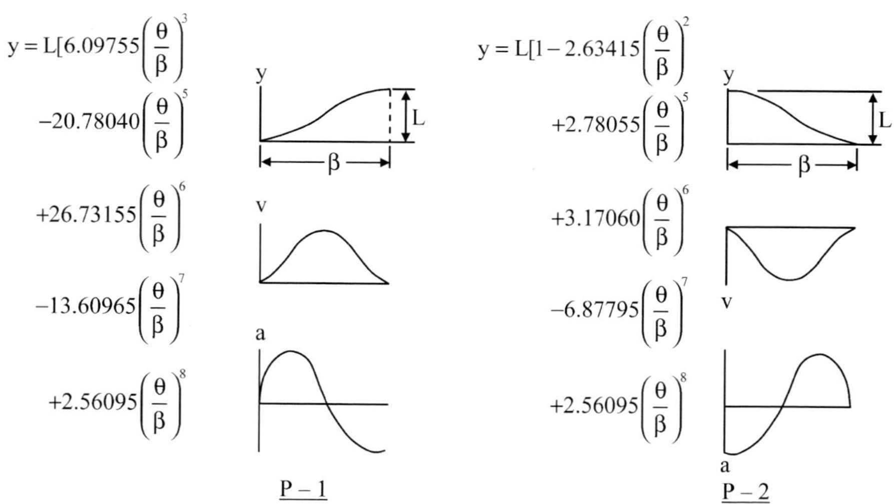

Fig. 5.15 Eighth-power polynomial characteristics

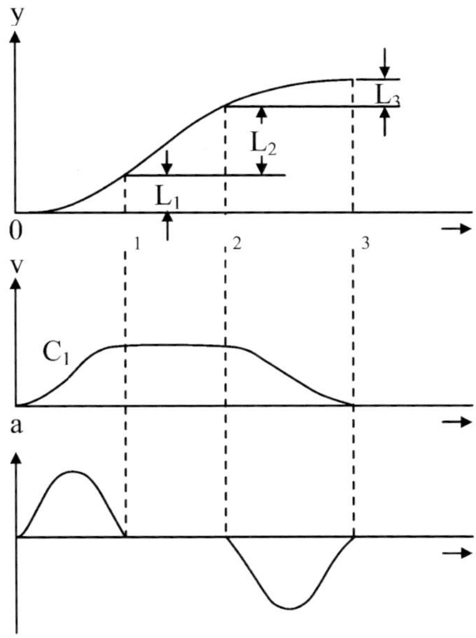

Fig. 5.16 Proposed solution

As another example compare maximum velocity, acceleration and jerks of constant acceleration, cosine acceleration, sine acceleration and their modified versions giving comments on each case.

Motion	Max. Vel	Max Accln.	Max. Jerk	Comments
Parabolic or Constant Accln.	$2L\omega/\beta$	$4L\omega^2/\beta^2$	∞	Jerk not acceptable
Trapezoidal Accln.	$2L\omega/\beta$	$5.3L\omega^2/\beta^2$	$44L\omega^3/\beta^3$	Not as good as modified trapezoid
Modified Trapezoidal Accln.	$2L\omega/\beta$	$4.888L\omega^2/\beta^2$	$61L\omega^3/\beta^3$	Low acceleration but rough jerk
Harmonic or Cosine Accln.	$1.571L\omega/\beta$	$4.945L\omega^2/\beta^2$	∞	Jerk not acceptable
Cycloidal or Sine Accln.	$2L\omega/\beta$	$6.283L\omega^2/\beta^2$	$40L\omega^3/\beta^3$	High acceleration, lower jerk but both are smooth
Modified Sine Accln.	$1.76L\omega/\beta$	$5.528L\omega^2/\beta^2$	$69L\omega^3/\beta^3$	Low velocity, moderate acceleration but higher jerk

5.3 Disk Cam with Knife-Edge Follower

In the graphical layout of the cam surface for a particular displacement of the follower, it is generally necessary to have as many divisions of the cam angle, so that the desired surface can be accurately obtained. One can use appropriate graphics software to do this as illustrated later in solved problems. However, in the examples below, only few divisions (six) are made for the purpose of illustration.

Minimum Size [Base Circle]

Cam surface with zero lift.

The procedure involves in holding the cam stationary and rotating the follower in a direction opposite to that of cam rotation.

The knife edge follower is rarely used in practice; however, it forms the basis of layout of cam surfaces. Let the follower motion be CV-1 for rise through 2 cm in 180° of cam rotation, dwell for 36° and then a return with CV-2 motion. The minimum size of the cam (base circle diameter) is 4 cm and it rotates in a counter-clockwise direction.

1. The first step is to draw the displacement diagram as shown in Figure 5.17a.
2. Draw the base circle 4 cm diameter and locate the starting point 0 as shown.
3. Divide the lift angle 180° into six equal divisions and mark these lines 1, 2, . . . , 6 in a direction opposite to that of the rotation of the cam, i.e., clockwise direction.
4. Project the displacements at positions 1, 2, . . . , 6 from the displacement diagram onto the initial position of the follower as shown.
5. Next, transfer these distances to the corresponding radial lines, by drawing arcs from the base circle center and draw a smooth curve through all these successive lift positions.
6. Draw an arc of 36° from the full rise position to the beginning of the return motion corresponding to the dwell.
7. Divide the fall angle 144° into six equal divisions and mark these lines 6, 7, . . . , 12 in clockwise direction beginning from the fall position (end of dwell).
8. Project the displacements at positions 6, 7, . . . , 12 from the displacement diagram onto the initial position of the follower.
9. Next, transfer these distances to the corresponding radial lines, by drawing arcs from the base circle center and draw a smooth curve through all these successive fall positions.

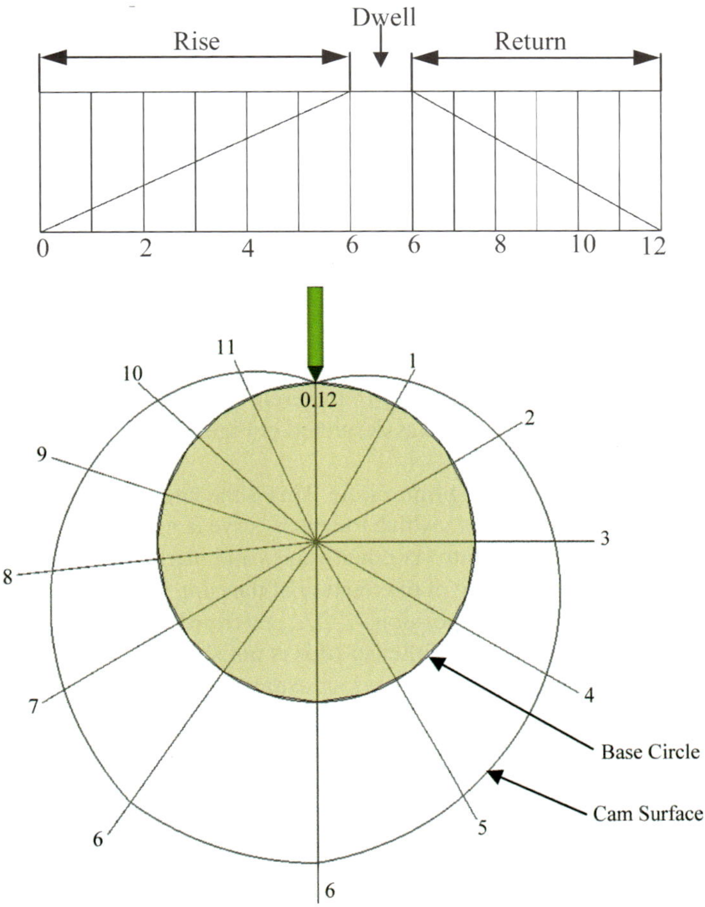

Fig. 5.17 (a) Displacement diagram. (b) Knife edge follower cam construction or base circle

5.4 Translating Roller Follower

Trace Point

It is a theoretical point on the cam follower corresponding to a knife-edge follower. This is used to generate a pitch curve.

Pitch Curve

This is a curve generated by the trace point. This is also the cam surface working with a knife-edge follower.

Prime circle

This is the minimum size of the cam through the pitch curve. Also, the base circle with a knife-edge follower.

As an example, motion PB-1 and PB-2 are used to take a roller translating follower through 2 cm lift and return with equal duration and no dwell. The minimum size of the cam is 4 cm and the roller diameter is 1 cm. The cam rotates in a counter-clockwise direction.

The graphical layout of the cam surface is shown in Figures 5.18a and b. Follow the steps given below:

1. Draw the base circle 4 cm diameter.
2. Draw the prime circle 2.5 cm radius (base circle radius + roller radius).
3. Draw the displacement diagram as described in Figure 5.10 for both rise and fall each through 180°.
4. Locate the trace point on the prime circle. This is the knife-edge point (the center of roller of the follower) from which the pitch curve is generated.
5. Divide the lift angle 180° into six equal angles and mark these lines 1, 2, ..., 6 in a direction opposite to that of the rotation of the cam, i.e., clockwise direction.
6. Project the displacements at positions 1, 2, ..., 6 from the displacement diagram onto the initial position of the follower (this is not shown to improve the clarity, these steps are the same as in the previous example).
7. Next, transfer these distances to the corresponding radial lines, by drawing arcs from the base circle center and draw a smooth curve through all these successive lift positions.
8. Divide the fall angle 180° into six equal divisions and mark these lines 6, 7, ..., 12 in clockwise direction beginning from the full rise position (dwell equal to zero).
9. Project the displacements at positions 6, 7, ..., 12 from the displacement diagram onto the initial position of the follower.
10. Next, transfer these distances to the corresponding radial lines, by drawing arcs from the base circle center, and draw a smooth curve through all these successive fall positions.
11. The curve obtained through steps 7 and 10 is the pitch curve.
12. From the pitch curve draw several circles of radius 0.5 cm (only a few are shown in the figure).
13. Draw a smooth curve tangent to all these circles to give the cam surface.

Pressure Angle

A pressure angle is the angle between the follower's motion and normal to the pitch curve (see Figure 5.18b). This angle gives the force direction of the cam and the roller of the follower. If the pressure angle is too large, the follower may get jammed in its bearings (recommended maximum pressure angle is 30°). Note that Simple

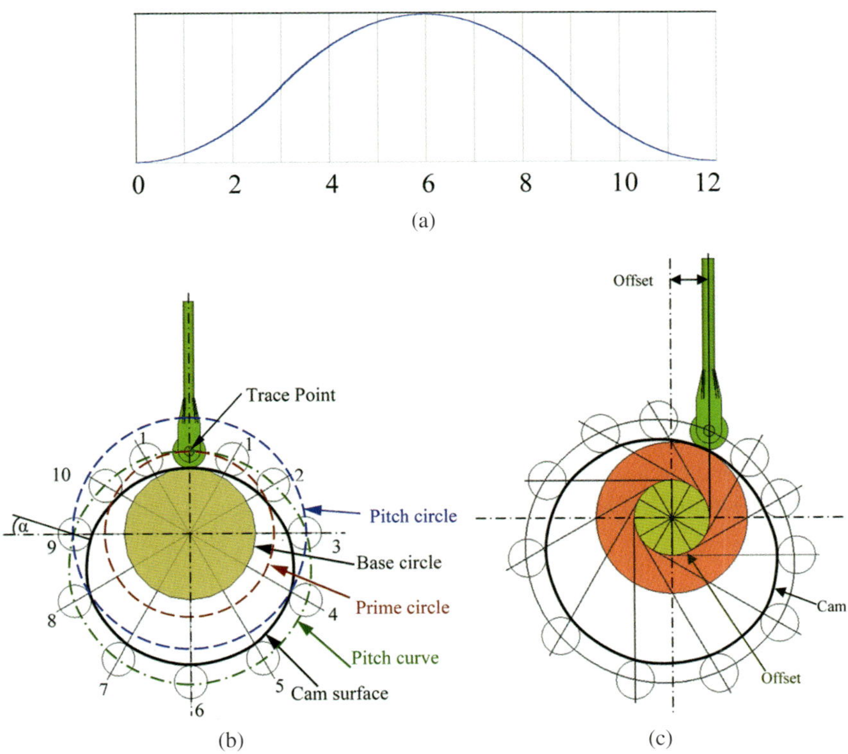

Fig. 5.18 (a) Displacement diagram. (b) Graphical layout of the cam. (c) Graphical layout of the cam with offset roller follower

harmonic motion gives a minimum pressure angle compared with cycloidal or polynomial motions.

Pitch Point

This is a point on the pitch curve where the maximum pressure angle occurs. For parabolic motion, maximum steepness occurs at the inflection points and the pressure angle is maximum here, also see Figure 5.18b.

Pitch Circle

This is a circle drawn through the pitch point with cam center.

Let the roller follower in a previous example be offset to the right of the cam center by 1 cm.

1. Draw the base circle.

2. Draw the offset circle as shown in Figure 5.18c.
3. Erect a vertical line to represent the translatory follower motion, which is tangential to the offset circle and to the right of the cam center.
4. Locate point 0 on this line, which is at a distance 2.5 cm (base circle radius + roller radius) from the center of the cam.
5. Divide the offset circle into the same number of divisions as the lift and fall motions are divided in the displacement diagram (six each at 30° intervals).
6. Draw lines tangential to the offset circle at these points to denote the translating roller follower position when the cam is held stationary and the follower is moved in a direction (clockwise) opposite to the direction of cam rotation and mark them with 1, 2, ..., 6 and 7, 8, ..., 12 as shown in the figure.
7. Next, transfer the displacements at the lift and fall positions to the corresponding radial lines, from the tangency points on the offset circle.
8. Draw a smooth curve through these points to give the pitch curve.
9. Draw several circles on this pitch curve with radius equal to 0.5 cm.
10. Draw a smooth curve tangent to all these circles to obtain the cam surface.

Analytical Design

Referring to Figure 5.19, define the trace point displacement (pitch curve) by

$$R = R_0 + f(\theta) \tag{5.41}$$

where $R_0 = R_c + R_r$ is the prime circle radius and $f(\theta)$ is follower displacement as a function of the cam angle of rotation θ. The radius of curvature at any point on the pitch curve is expressed as

$$\rho = \frac{\left[R^2 + \left(\frac{dR}{d\theta}\right)^2\right]^{1.5}}{R^2 + 2\left(\frac{dR}{d\theta}\right)^2 - R\frac{d^2R}{d\theta^2}} \tag{5.42}$$

Pointing [Undercutting]

Whenever the radius of curvature is less than the radius of the follower, pointing occurs. Figure 5.20a shows the case when the radius of curvature is larger than the radius of the roller of the follower. When the radius of curvature of the pitch curve is equal to the roller radius, we have a limiting case, as shown in Figure 5.20b. In Figure 5.20c, the radius of curvature of the pitch curve is less than the roller radius. In this case, the cam misses some of the material which should be otherwise there and the follower does not properly replicate the desired motion.

To avoid pointing or undercutting of the cam, the minimum radius of the pitch curve in a complete revolution of the cam should be larger than the radius of the roller follower. Equation (5.42) can be used to determine the radius of curvature at

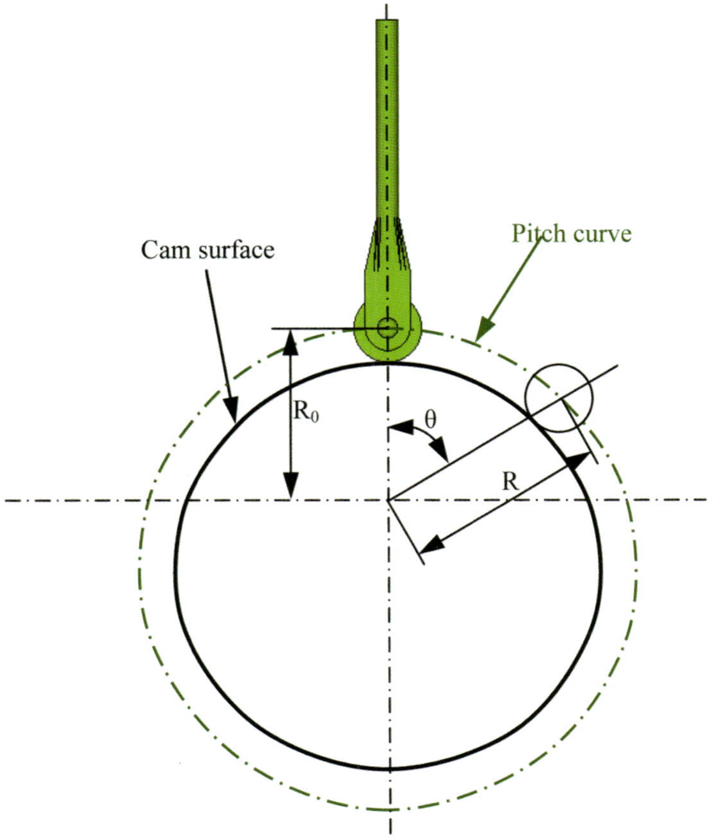

Fig. 5.19 Analytical design of cams

various positions of the cam in a complete revolution, 0 to 360°, and determine the minimum radius of curvature, $\rho_{\min}$, and its location. For no pointing or undercutting, $\rho_{\min} \geq R_r$.

For a given $R(\theta)$, equation (5.42) can be differentiated with respect to θ and thus can find the location $\theta_{\min}$ where $\rho_{\min}$ occurs. Upon substituting the value of $\theta_{\min}$ in equation (5.42), $\rho_{\min}$ can be determined. As this is not a simple mathematical procedure, the minimum values of the radius of curvature for different follower motions are given graphically in Figures 5.21 through 5.23 for different follower motions. These figures can be used in the design of a translating roller follower and ensure that the cam is not pointed.

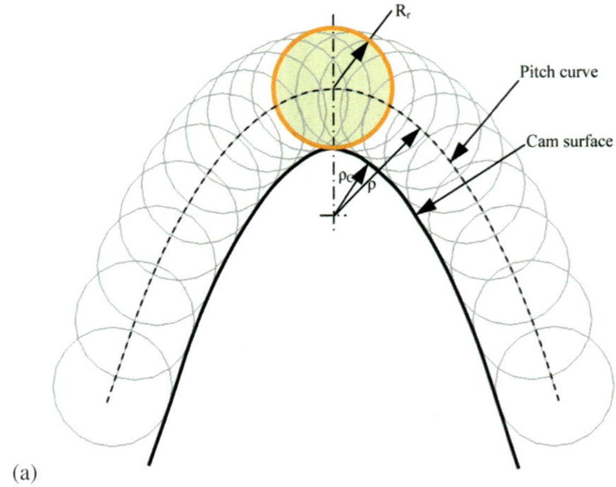

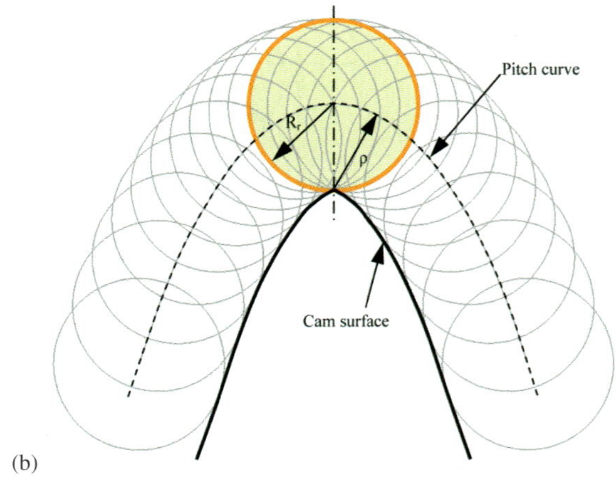

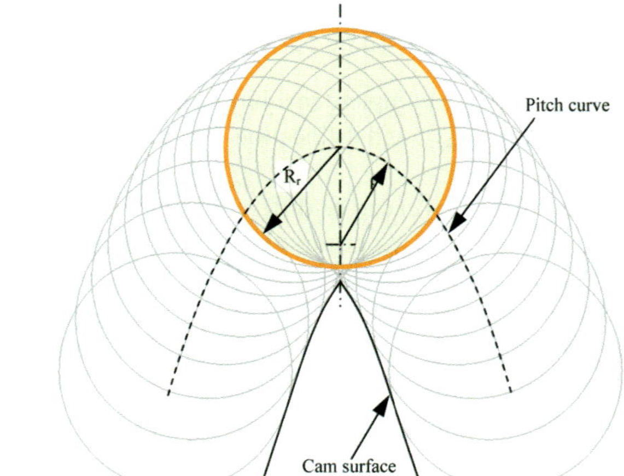

Fig. 5.20 (a) Radius of curvature larger than radius of the roller of the follower. (b) Radius of curvature equal to radius of the roller of the follower – limiting case. (c) Radius of curvature less than radius of the roller of the follower – pointing occurs.

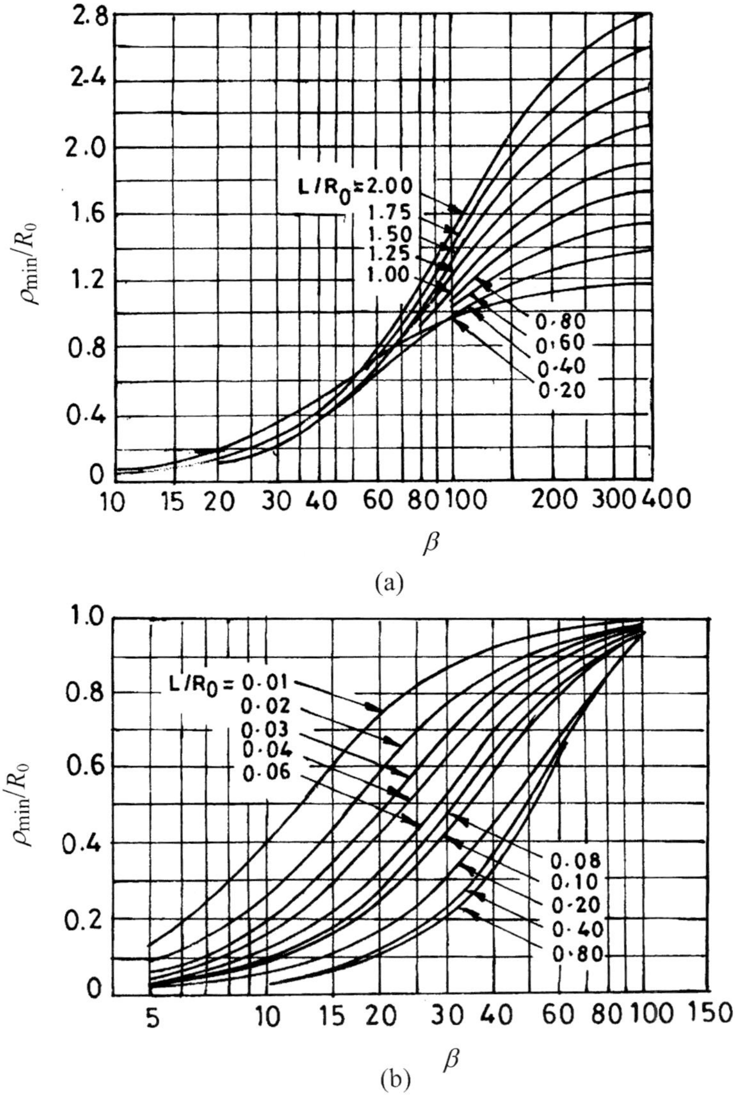

Fig. 5.21 ρ_{min} for simple harmonic motion as a function of active cam angle β

Maximum Pressure Angle

As mentioned before, the pressure angle should be kept as low as possible to avoid jamming of the follower in its bearings. Therefore, the maximum pressure angle for a given motion should be determined and this should be lower than a specified value.

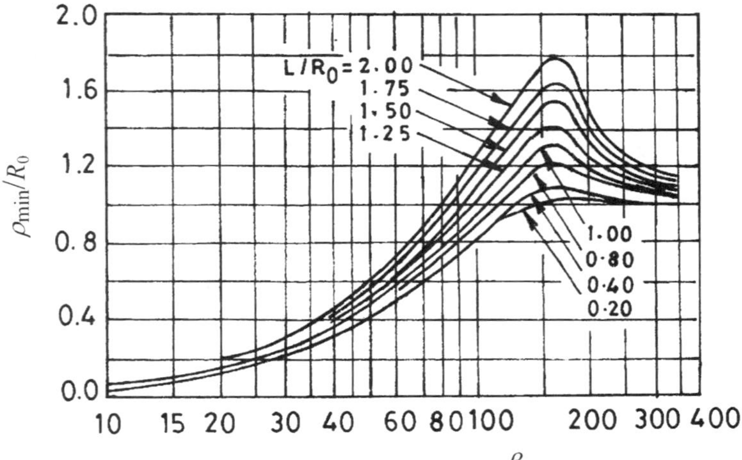

(a)

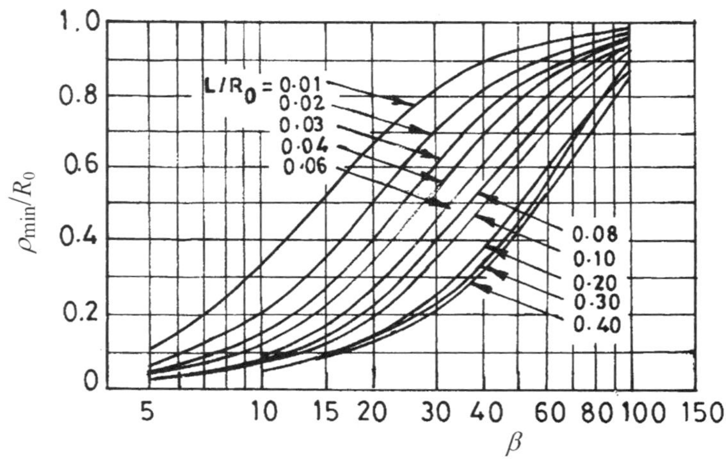

(b)

Fig. 5.22 $\rho_{\min}$ for cycloid motion as a function of active cam angle β.

Refer to Figure 5.24; α represents the pressure angle for the position PQ of the roller follower. For a small counter-clockwise rotation of the cam, i.e., clockwise rotation of the follower with the cam held stationary, let the roller move to $P'Q'$ from PQ. Then the angle α' is

$$\alpha' = \tan^{-1}\frac{P'M}{PM} \tag{5.43}$$

Also,

$$\angle OPM_{\Delta\theta\to 0} = 90^{\circ} \quad \text{and} \quad \angle QPP'_{\Delta\theta\to 0} = 90^{\circ} \tag{5.44}$$

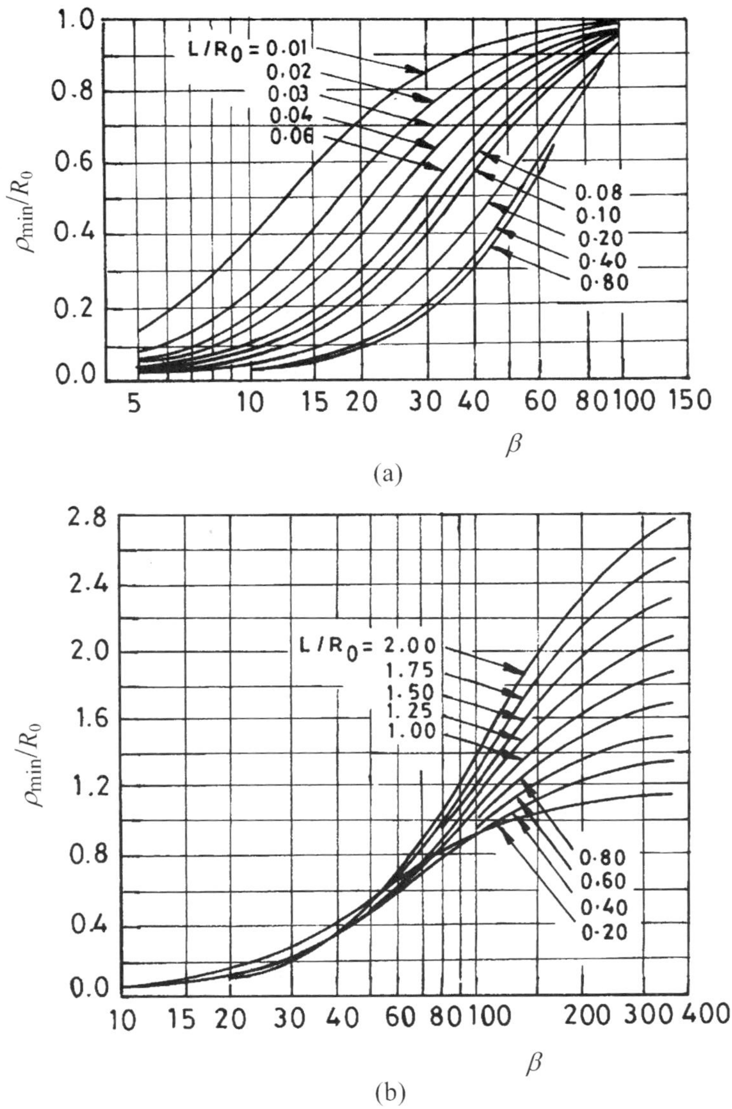

Fig. 5.23 $\rho_{\min}$ for eighth-power polynomial motion as a function of active cam angle β.

That is, *PL* merges with *PN* as $\Delta\theta \to 0$, both approaching $PM = R\Delta\theta$. Therefore,

$$\alpha = \lim_{\Delta\theta \to 0} \alpha' = \tan^{-1} \frac{1}{R}\frac{dR}{d\theta} \tag{5.45}$$

Solving the above equation for a given $R(\theta)$, the pressure angle can be determined for various values of θ and the maximum value can also be determined. This is again not an easy procedure and therefore, a nomogram devised by Varnum, is provided

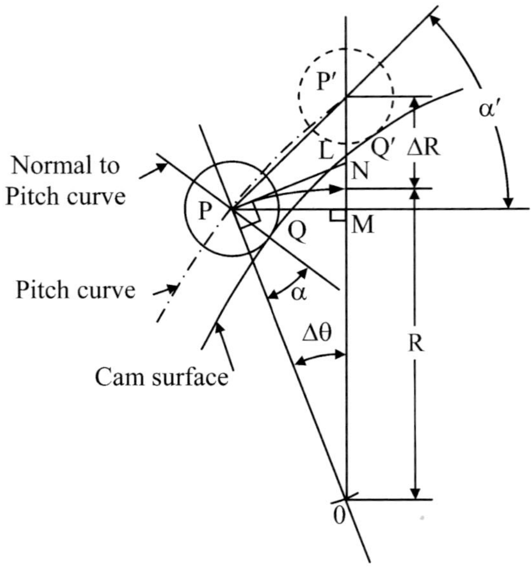

Fig. 5.24 Determination of the maximum pressure angle

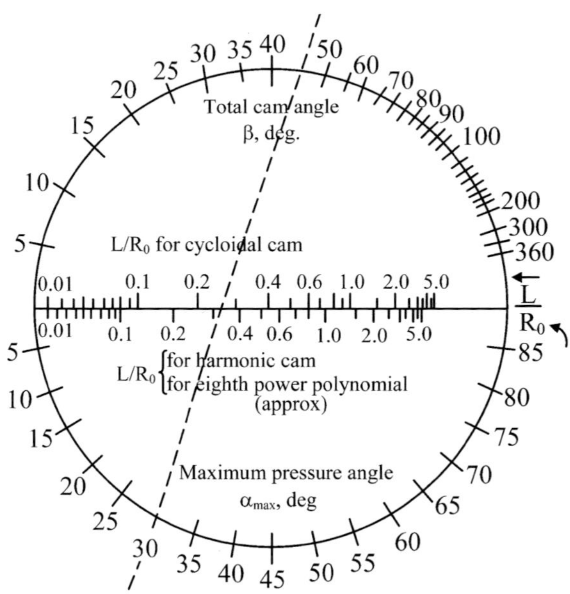

Fig. 5.25 Varnum's nomogram

in Figure 5.25 to determine the maximum pressure angle for three different follower motions.

As an example let a radial roller follower be required to have a total displacement of 2 cm with cycloidal motion for a cam angle β of 60°. Determine the minimum radius of the cam and the maximum roller follower radius to limit the maximum pressure angle to 30°, and suggest suitable values.

1. From Figure 5.25, for $\beta = 60°$ and $\alpha_{\max} = 30°$, $L/R_o = 0.35$.
2. Therefore, $R_o = 2/0.35 = 5.72$ cm.
3. From Figure 5.22, for $\beta = 60°$ and $L/R_o = 0.35$, $\rho_{\min}/R_o = 0.55$.
4. Hence, $\rho_{\min} = 0.55 \times 5.72 = 3.146$ cm.
5. Therefore, we can choose $R_o = 5.75$ cm and $R_r = 3.0$ cm.

5.5 Translating Flat Follower

Graphical Layout

First we consider the graphical layout of a cam with a translating flat follower.

Draw the cam profile for a translating flat follower to rise and fall through 1.4 cm without any dwell in equal intervals. The minimum size of the cam is 4 cm; Use simple harmonic motion.

1. Draw the displacement diagram with six equal intervals for the rise and fall motions as shown in Figure 5.26.
2. Draw the minimum size of the cam with radius equal to 2 cm.
3. Divide the base circle into the same number of parts for both the rise and fall through 180° as the displacement diagram and mark the radial lines 1, 2, ..., 12.
4. Project the displacements at positions 1, 2, ..., 12 from the displacement diagram onto the initial position of the follower. (This is not shown to improve the clarity; these steps are same as in previous example.)
5. Next, transfer these distances to the corresponding radial lines, by drawing arcs from the base circle center.
6. Draw lines normal to these radial lines at the corresponding lift positions to represent the flat follower positions.
7. Draw a smooth curve that is tangential to all the flat follower lines (do not join these lift positions by lines or a smooth curve) to give the cam surface.
8. Note the position where the contact between the cam surface and the follower is the farthest from the corresponding radial line; let this be W. Then the width of the flat follower required is $2W$.

Analytical Design

Consider Figure 5.27. The contact point denoted by (x, y) is located at a distance w from the center line of the follower. The displacement of the follower from the center of cam is

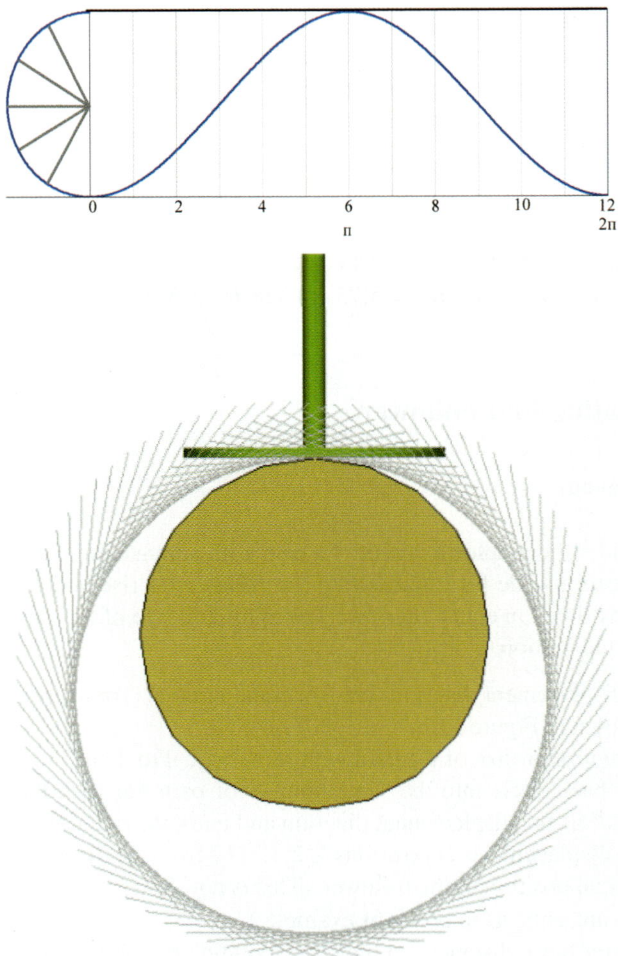

Fig. 5.26 Graphical layout of translating flat follower

$$R = R_c + f(\theta) \tag{5.46}$$

From Figure 5.27,

$$\frac{dy}{dx} = -\cot\theta$$

i.e.

$$\frac{dy}{d\theta}\sin\theta = -\frac{dx}{d\theta}\cos\theta \tag{5.47}$$

and

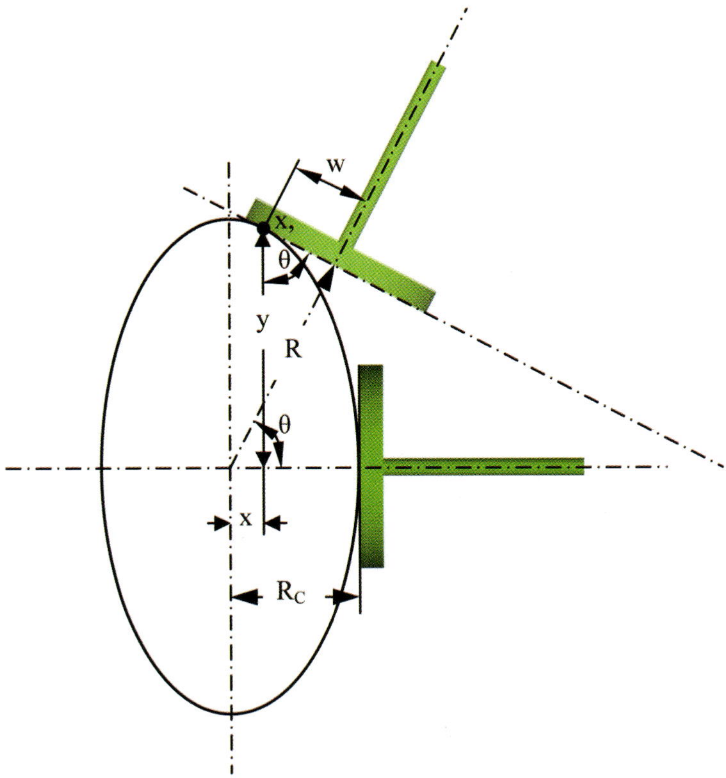

Fig. 5.27 Analytical design of a cam with translating flat follower

$$\begin{aligned} R &= y\sin\theta + x\cos\theta \\ w &= y\cos\theta - x\sin\theta \end{aligned} \tag{5.48}$$

From the above equation (5.48), we find

$$\begin{aligned} \frac{dR}{d\theta} &= y\cos\theta - x\sin\theta \\ w &= \frac{dR}{d\theta} = f'(\theta) \end{aligned} \tag{5.49}$$

and solving the equations in (5.48), we get the cam surface

$$\begin{aligned} x &= \left[R_c + f(\theta)\right]\cos\theta - f'(\theta)\sin\theta \\ y &= \left[R_c + f(\theta)\right]\sin\theta + f'(\theta)\cos\theta \end{aligned} \tag{5.50}$$

Cusp

A cusp is a point on the cam surface wherein the follower experiences no displacement even though the cam has a rotation.

Such a condition of the cusp is illustrated in Figure 5.28, wherein the follower at position (x, y) suffers no displacement, even though the cam rotates through an angle $d\theta$ from position θ. This condition can be expressed mathematically by

$$\frac{dx}{d\theta} = \frac{dy}{d\theta} = 0 \tag{5.51}$$

Obviously such a condition where a cam surface may have a cusp should be avoided. The condition to avoid cusps on the cam surface with a flat follower can be obtained as follows. Differentiate equation (5.50) with respect to θ to give

$$\begin{aligned}\frac{dx}{d\theta} &= -\left[R_c + f(\theta) + f''(\theta)\right]\sin\theta \\ \frac{dy}{d\theta} &= \left[R_c + f(\theta) + f''(\theta)\right]\cos\theta\end{aligned} \tag{5.52}$$

Equation (5.51) will be true only when

$$R_c + f(\theta) + f''(\theta) = 0$$

Therefore, to avoid cusps,

$$R_c + f(\theta) + f''(\theta) > 0 \tag{5.53}$$

As an example determine the base circle diameter of a disk cam and the width of the translating flat follower to have a total lift of 4 cm during 90° of rotation and return with simple harmonic motion during 90°, with two equal dwells.

$$\begin{aligned}f(\theta) &= \frac{1}{2}L\left(1 - \cos\frac{\pi\theta}{\beta}\right) = 2(1 - \cos 2\theta) \\ f'(\theta) &= 4\sin 2\theta \\ f''(\theta) &= 8\cos 2\theta\end{aligned}$$

To avoid cusps, $R_c + f(\theta) + f''(\theta) > 0$, i.e., $R_c + 2 + 6\cos 2\theta > 0$.

Since the minimum value of $\cos 2\theta = -1$, $R_c > 4$ cm. Let $R_c = 4.5$ cm. The width of the follower is given by twice that of the maximum value of $w = dR/d\theta = f'(\theta)$ (because of symmetry).

Since $f'(\theta)_{\max} = [4\sin 2\theta]_{\max} = -4$, the follower width can be taken as 8.5 cm. The cam surface is plotted from equation (5.50)

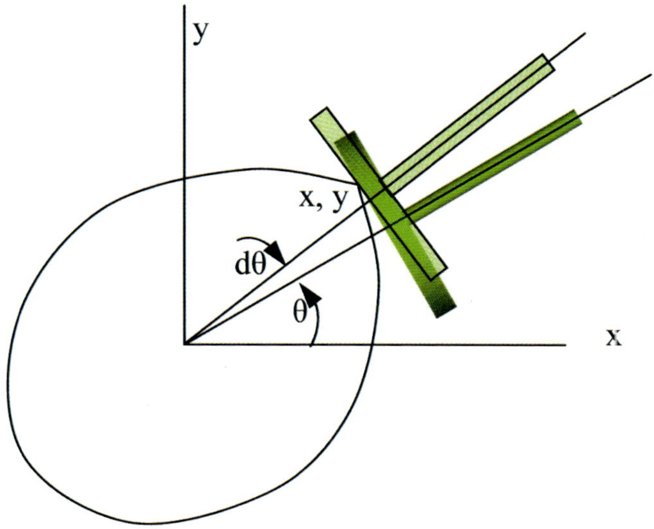

Fig. 5.28 Condition for a cusp

$$x = [6.5 - 2\cos 2\theta]\cos\theta - 4\sin 2\theta \sin\theta$$

$$y = [6.5 - 2\cos 2\theta]\sin\theta + 4\sin 2\theta \cos\theta$$

5.6 Oscillating Flat Follower

The minimum size of a cam to operate an oscillating flat follower is 10 cm. When the flat follower is horizontal, its center of rotation is at a radius of 15 cm, making an angle of 30° with a horizontal line passing through the cam center. The oscillating follower is required to move through a 20° angle with an arm 15 cm in length. The lift motion for the rise is Parabolic with a cam angle of 180° following immediately with a simple harmonic motion with a cam angle of 120°. The cam rotation is counter-clockwise.

- Draw a circle representing the minimum size of the cam as shown in Figure 5.29.
- Locate the center of the oscillating follower on the line making 30° with the horizontal line at the cam center.
- Draw a horizontal line representing the flat surface of the follower in the zero lift position.
- Draw a circle with the follower center of oscillation as the center and tangential to the flat follower line in the zero lift position.
- Locate point G and mark GC as representing the follower arm.

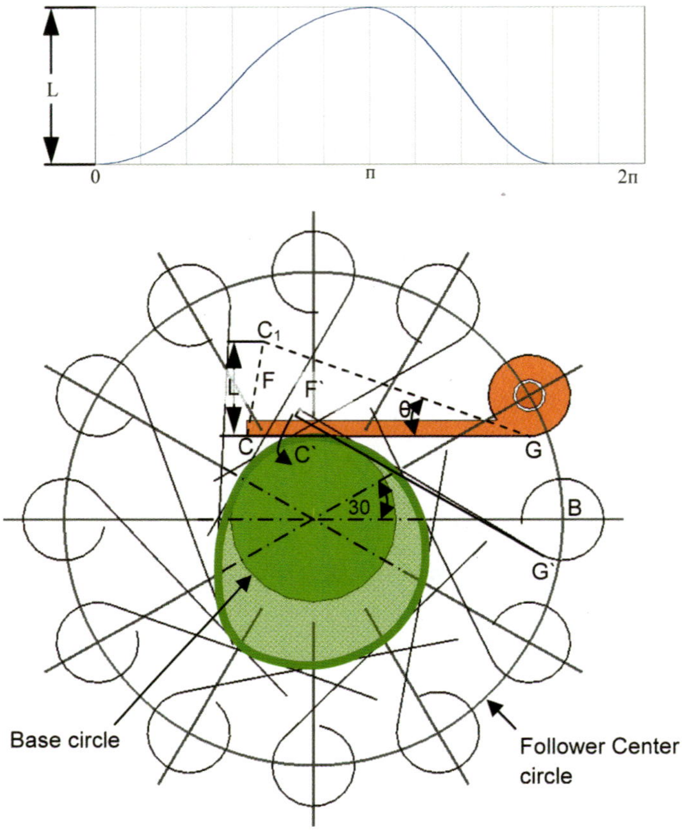

Fig. 5.29 Oscillating flat follower

- The linear lift length is $15 \times 20 \times \pi/180 = 5.236$ cm. Construct the displacement diagram as shown in Figure 5.29 (details not given here). Only six divisions for the lift and four divisions for the fall motion are shown for clarity.
- The lift at different locations should be converted to arc lengths while drawing the cam surface.
- As usual keep the cam stationary and rotate the follower around in the opposite direction of the cam rotation. The center of the follower for the first division is marked B'.
- For the first lift division, identify the $G'C'$ for no lift. Mark the arc length CF on CC_1 and transfer this to the first division $C'F'$, to give $G'C'F'$. Draw the follower position at the first division of the lift, $G'F'$.
- Repeat the above step and locate the follower positions at all lift and fall positions.
- Draw a smooth curve tangential to all the follower positions to obtain the cam surface.

If you have an oscillating roller follower, first generate the pitch curve using the trace point and then obtain the cam surface.

5.7 Cams of Specified Contour

Manufacturing of cams with simple contours is inexpensive and therefore they are preferred in such applications where the cam follower path does not have to follow a specified route.

Eccentric Circle Cam with Translating Flat Follower

This is a circular cam with the center of rotation offset from its geometric center. Figure 5.30a shows such a cam with eccentricity E. Let the displacement of the follower be y for a cam angle of rotation θ. This motion can also be generated identically by an equivalent Scotch–Yoke mechanism shown in Figure 5.30b, in which the displacement $y_\varepsilon = y$ is obtained for the crank rotation angle $\varepsilon = \theta$. Note the starting positions in both the cam-follower and the Scotch–Yoke mechanisms.

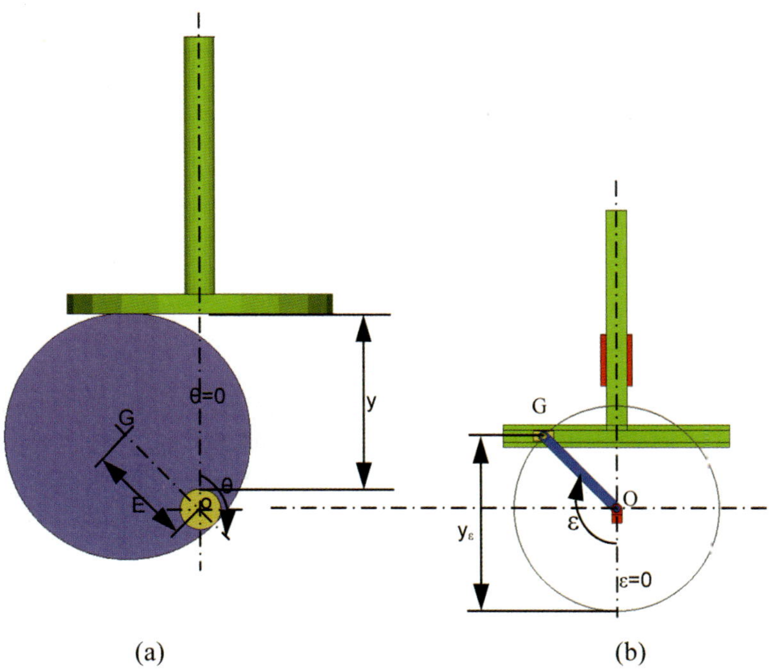

Fig. 5.30 Eccentric circle cam with translating flat follower

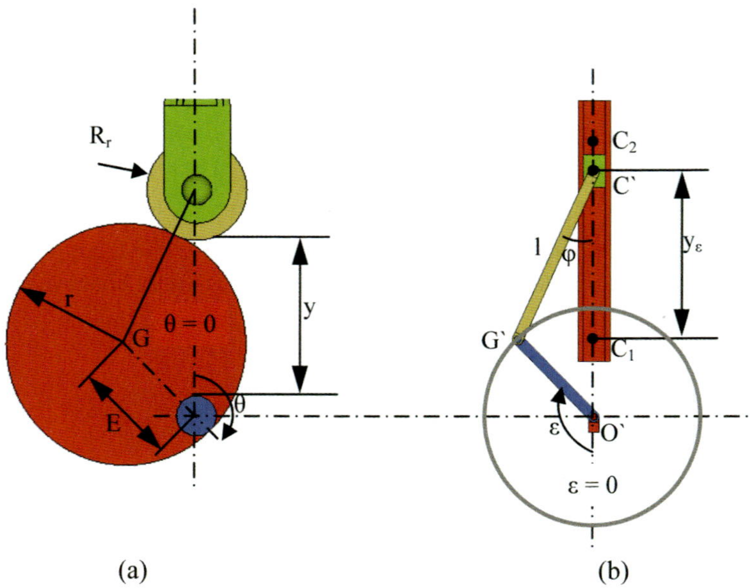

Fig. 5.31 Eccentric circle cam with translating roller follower

The displacement of the follower is

$$y = E\,(1 - \cos\theta) \tag{5.54}$$

The corresponding velocity and accelerations are

$$\begin{aligned} v &= E\omega\sin\theta \\ a &= E\omega^2\cos\theta \end{aligned} \tag{5.55}$$

Eccentric Circle Cam with Translating Roller Follower

This cam and its equivalent reciprocating linkage are shown in Figure 5.31. In Figure 5.31b, the crank radius is E and the connecting rod length is l. From the starting position $\varepsilon = 0$, with the slider at lower dead center C_1, the displacement of the slider is $y_\varepsilon = y$ for $\varepsilon = \theta$. We note that

$$l = r + R_r \tag{5.56}$$

The lift expression can be obtained from

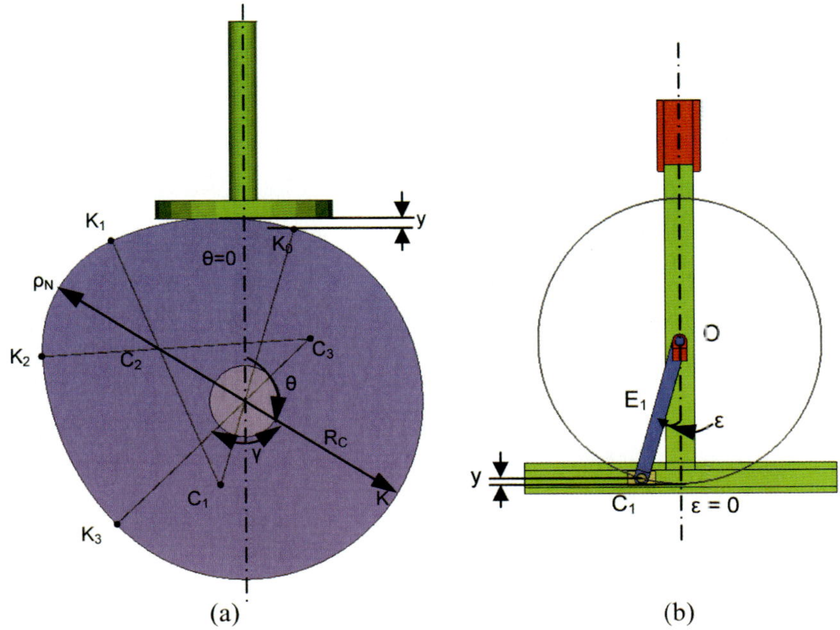

Fig. 5.32 Circular arc cam [triple-curve] with translating flat follower (K_0K_1)

$$y = OC' - OC_1 = [E\cos(180-\theta) + l\cos\varphi] - [l - E]$$
$$= E - l - E\cos\theta + l\cos\varphi$$

Since, $\sin\varphi = (\pi/l)\sin\theta$ we can rewrite the above equation as

$$y = E\,(1-\cos\theta) - l + \sqrt{l^2 - E^2\sin^2\theta} \tag{5.57}$$

Circular Arc Cam [Triple-Curve] with Translating Flat Follower

This cam has three circular arcs K_0K_1 (rise), K_1K_2 (rise and fall equally divided) and K_2K_3 (fall) having centers C_1, C_2 and C_3 respectively as shown in Figure 5.32a. The base circle radius is R_c and has a dwell period 2γ marked by the region K_3KK_0. In general C_1OK_0, C_2OK, C_3OK_3, $C_1C_2K_1$ and $C_3C_2K_2$ are all made to lie on different lines as shown. It is also conventional to make $OC_1 = OC_2 = OC_3$.

The equivalent Scotch–Yoke for the rise motion on K_0K_1 having the geometric center at C_1 and center of rotation at O is shown in Figure 5.32b. Here,

$$\varepsilon = \theta - \gamma \quad \text{and} \quad y = y_\varepsilon \tag{5.58}$$

Let the eccentricity of the first curve K_0K_1 given by OC_1 be equal tc E_1. Then

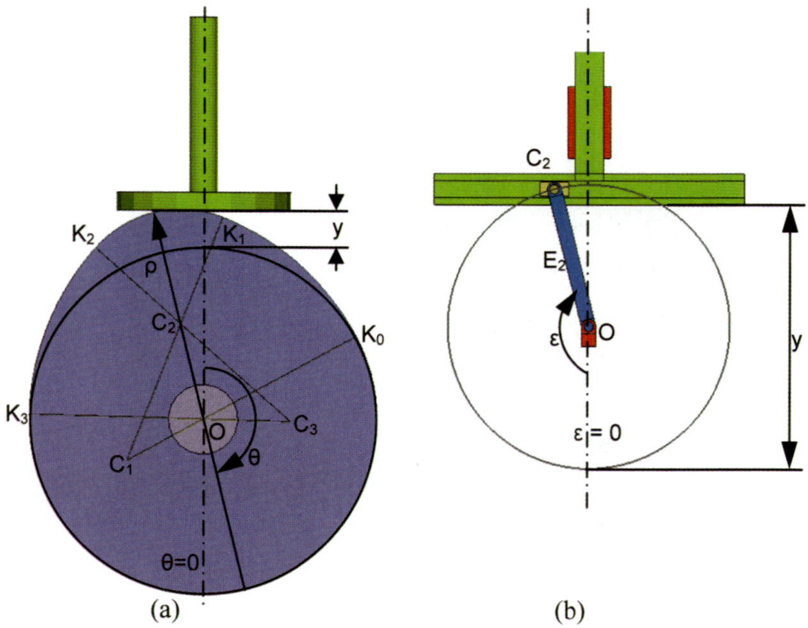

Fig. 5.33 Circular arc cam [triple-curve] with translating flat follower (K_1K_2)

$$y = E_1\,(1 - \cos\varepsilon) \tag{5.59}$$

The equivalent Scotch–Yoke for the motion on K_1K_2 is given in Figure 5.33. It is evident that

$$\varepsilon = \theta \quad \text{and} \quad y = y_\varepsilon \tag{5.60}$$

Let

$$y = y_\varepsilon + X \tag{5.61}$$

To determine X, consider the full lift position, where

$$L = 2E_2 + X \tag{5.62}$$

The eccentricity for this arc and nose radius ρ_N from Figure 5.33, are related by

$$E_2 = L + R_c - \rho_N \tag{5.63}$$

From the above two equations (5.62) and (5.63), we get

$$X = 2\rho_N - 2R_c - L \tag{5.64}$$

Therefore,

$$y = E_2\,(1 - \cos\varepsilon) + 2\rho_N - 2R_c - L \tag{5.65}$$

For the arc K_2K_3 equations (5.58) and (5.59) will hold.

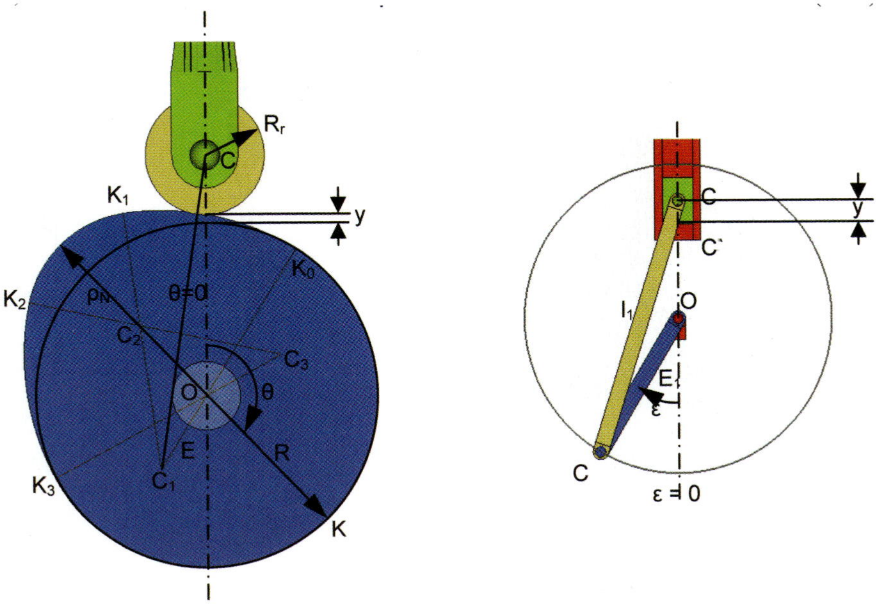

Fig. 5.34 Circular arc cam [triple-curve] with translating roller follower for the arc K_0K_1

Circular Arc Cam [Triple-curve] with Translating Roller Follower

For the arc K_0K_1, the equivalent slider crank mechanism is shown in Figure 5.34. The crank radius is E_1 and the connecting rod length $l_1 = E_1 - R_c + R_r$. It is evident that

$$\varepsilon = \theta - \gamma \quad \text{and} \quad y = y_\varepsilon \tag{5.66}$$

$$y = E_1 - l_1 - E_1 \cos(\theta - \gamma) + \sqrt{{l_1}^2 - {E_1}^2 \sin^2(\theta - \gamma)} \tag{5.67}$$

For the roller follower acting on the nose of the cam, the equivalent linkage is shown in Figure 5.35. The crank radius is E_2 and the connecting rod length is $l_2 = \rho_N - R_r$. It is evident that

$$\varepsilon = \theta \quad \text{and} \quad y = y_\varepsilon \tag{5.68}$$

Expressing y in a similar manner as in the case of translating flat follower, the value of X in the equation $y = y_\varepsilon - X$ can be determined as $X = 2\rho_N - R_c - L$, see equation (5.64). Hence,

$$y = E_2\,(1 - \cos\varepsilon) - l_2\,(1 - \cos\varphi) + 2\rho_N - 2R_c - L \tag{5.69}$$

Using $\sin\varphi = (\pi/l)\sin\theta$ the above equation can be written as

$$y = E_2\,(1 - \cos\theta) - l_2 + \sqrt{{l_2}^2 - {E_2}^2 \sin^2\theta} + 2\rho_N - 2R_c - L \tag{5.70}$$

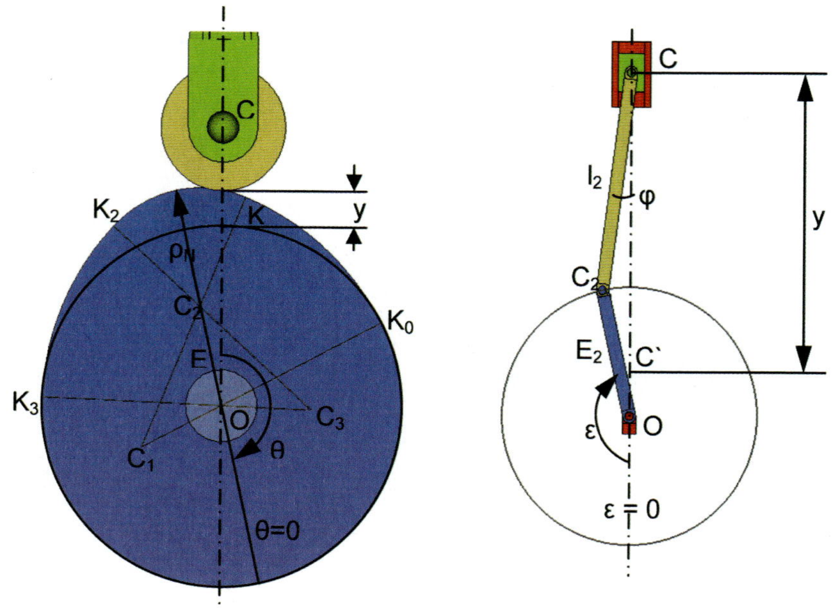

Fig. 5.35 Circular arc cam [triple-curve] with translating roller follower for the nose

Tangent Cam with Roller Follower

This is a triple curve cam in which the arcs K_0K_1 (rise) and K_2K_3 (fall) have infinite radius, i.e., the two flanks now become straight lines tangential to the base and nose circles. To determine the lift on the first straight line flank, consider the triangles OK_0I and IPC in Figure 5.36 from which we get

$$OI = R_c \sec\theta$$
$$IC = R_r \sec\theta \tag{5.71}$$

Also,

$$y = OC - R_c - R_r \tag{5.72}$$

Therefore

$$y = (R_c + R_r)(\sec\theta - 1) \tag{5.73}$$

Differentiating once with respect to time, the velocity is given by

$$v = \omega\,(R_c + R_r)\sec\theta\tan\theta \tag{5.74}$$

The acceleration is given by one more differentiation

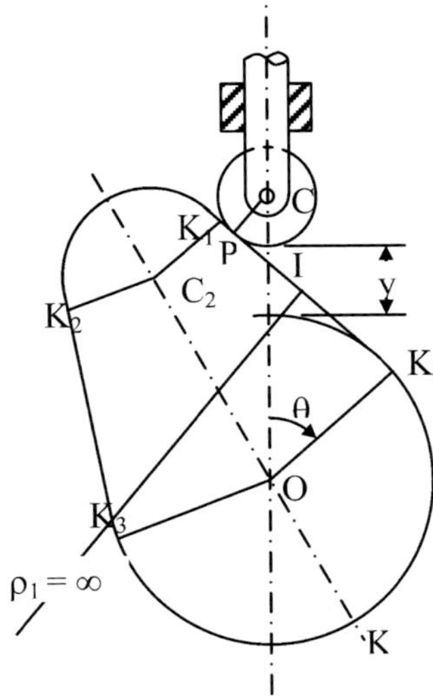

Fig. 5.36 Tangent cam with roller follower

$$a = \omega^2 (R_c + R_r) \left[\sec\theta \tan^2\theta + \sec\theta (1 + \tan^2\theta)\right]$$

$$= \omega^2 (R_c + R_r) \frac{1 + \sin^2\theta}{\cos^3\theta} \quad (5.75)$$

The action of the roller follower on the nose of the tangent cam is the same as that of the triple curve circular arc cam.

5.8 Solved Problems

Solved Problem 5.1

Construct the profile of a disk cam with a translating roller follower. Base circle dia 15 cm, roller dia 1.5 cm, lift 5 cm in 180°, return in 135° and then dwell. Both the lift and return motions are to be harmonic. The cam rotates in a clockwise direction. Check whether there will be undercutting on the cam profile. What is the maximum pressure angle?

The displacement, velocity and acceleration of the rise and return motions in 10° intervals are calculated from (5.25) and tabulated below:

	θ	y	v	a
0	0	0.00000	0.00000	2.5000
1	10	0.03798	0.43412	2.4620
2	20	0.15077	0.85505	2.3492
3	30	0.33494	1.25000	2.1651
4	40	0.58489	1.60697	1.9151
5	50	0.89303	1.91511	1.6070
6	60	1.25000	2.16506	1.2500
7	70	1.64495	2.34923	0.8551
8	80	2.06588	2.46202	0.4341
9	90	2.50000	2.50000	0.0000
10	100	2.93412	2.46202	−0.4341
11	110	3.35505	2.34923	−0.8551
12	120	3.75000	2.16506	−1.2500
13	130	4.10697	1.91511	−1.6070
14	140	4.41511	1.60697	−1.9151
15	150	4.66506	1.25000	−2.1651
16	160	4.84923	0.85505	−2.3492
17	170	4.96202	0.43412	−2.4620
18	180	5.00000	0.00000	−2.5000
19	190	4.93261	−0.76872	−4.3246
20	200	4.73408	−1.49600	−3.9717
21	210	4.41511	−2.14263	−3.4046
22	220	3.99290	−2.67374	−2.6540
23	230	3.49020	−3.06072	−1.7603
24	240	2.93412	−3.28269	−0.7718
25	250	2.35464	−3.32769	0.2584
26	260	1.78299	−3.19330	1.2747
27	270	1.25000	−2.88675	2.2222
28	280	0.78440	−2.42458	3.0500
29	290	0.41128	−1.83170	3.7133
30	300	0.15077	−1.14007	4.1764
31	310	0.01690	−0.38698	4.4144
32	315	0.00000	0.00000	4.4444

Figure 5.37 gives the construction of the cam profile. Note that the follower is taken counter-clockwise around the cam.

1. Draw the base circle 15 cm diameter.
2. Draw the prime circle 8.25 cm radius (base circle radius + roller radius).
3. Locate the trace point on the prime circle. This is the knife-edge point (the center of roller of the follower) from which the pitch curve is generated.
4. Divide the lift angle 180° into 18 equal divisions and mark these lines 1, 2, ..., 18 (not shown in figure) in a direction opposite to that of the rotation of the cam, i.e., counter clockwise direction.
5. Transfer the displacements at positions 1, 2, ..., 18 from the table above onto the corresponding radial lines from the prime circle and draw a smooth curve through all these successive lift positions to get the pitch curve.
6. Divide the fall angle 135° into 13 equal divisions of 10° and the last division of 5° corresponding to the positions given in table above. Mark these lines 19, 20,

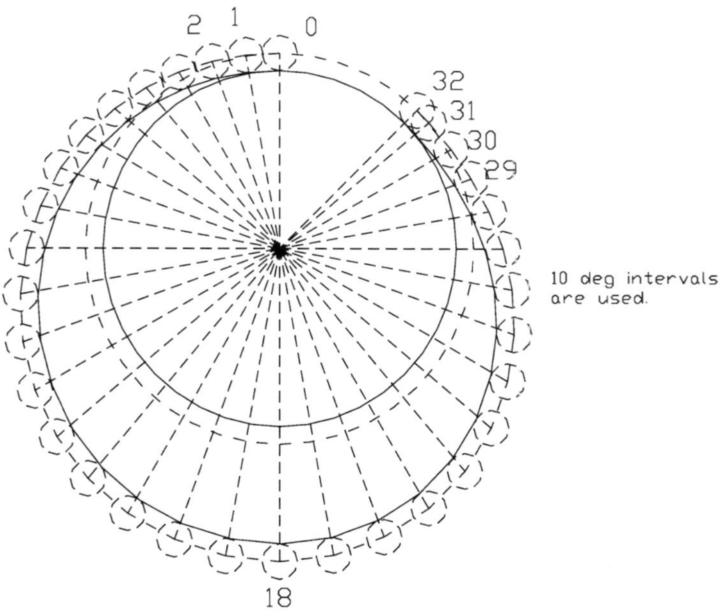

Fig. 5.37

..., 31, 32 in counter-clockwise direction beginning from the full rise position (dwell equal to zero).

7. Transfer the displacements at positions 19, 20, ..., 32 from the table above onto the corresponding radial lines from the prime circle and draw a smooth curve through all these successive lift positions to get the pitch curve in the return motion. For the rest of the cam angle 45°, in the dwell portion, the pitch curve is the same as the prime circle.
8. From the pitch curve draw several circles of radius 0.75 cm.
9. Draw a smooth curve tangent to all these circles to give the cam surface.

1. Check for undercutting in the rise portion: $L/R_0 = 5/8.25 = 0.606$.
2. From Figure 5.21a, $\rho_{min}/R_0 = 1.35$ for $\beta = 180°$, $\rho_{min} = 11.1375 > 0.75$.
3. Hence no undercutting occurs in the rise portion.
4. Check for undercutting in the return portion.
5. From Figure 5.21a, $\rho_{min}/R_0 = 1.20$ for $\beta = 135°$, $\rho_{min} = 9.9 > 0.75$.
6. Hence no undercutting occurs in the return motion also.
7. From Figure 5.25, $\alpha_{max} = 18°$ for $\beta = 135°$ and $\alpha_{max} = 15°$ for $\beta = 180°$. Maximum pressure angle is 18°.

Solved Problem 5.2

In the above problem, the roller follower is offset by 2 cm and the lift portion is cycloidal while the fall portion is parabolic. Construct the cam profile.

1. First construct the base circle and then the offset circle, see Figure 5.38.
2. Locate the trace point on the vertical line from zero position on the offset circle, such that, the roller will touch the base circle at this position. Find the trace point height which is 8.310385 cm.
3. Determine the lift displacements at conveniently divided number of cam angle positions, here 12 in number at 15° intervals, see the table given below. (If you prefer you can follow the graphical construction.)
4. Find the trace point distances at each of these locations from the offset positions of the follower (tangential to the offset circle), see table given below. Locate these points and join them to give a smooth curve (pitch curve).
5. Repeat the above two steps 3 and 4 for the return motion. (Use 12 steps for the parabolic motion at 11.25° intervals.)
6. Complete the pitch curve by drawing a circle in the dwell period.
7. Draw several circles of roller radius at each of the pitch curve positions and draw the cam surface which is tangential to all these rollers.

Cycloidal motion lift

$$y = L\left(\frac{\theta}{\beta} - \frac{1}{2\pi}\sin\frac{2\pi\theta}{\beta}\right)$$

θ	y	Trace point	
0	0	0.0000	8.3104
1	15	0.0188	8.3292
2	30	0.1441	8.4545
3	45	0.4543	8.7646
4	60	0.9775	9.2879
5	75	1.6854	9.9958
6	90	2.5000	10.8104
7	105	3.3146	11.6249
8	120	4.0225	12.3329
9	135	4.5458	12.8562
10	150	4.8558	13.1662
11	165	4.9812	13.2916
12	180	5.0000	13.3104

Parabolic motion return

$$\text{(First half)} \quad y = L\left[1 - 2\left(1 - \frac{\theta}{\beta}\right)^2\right]$$

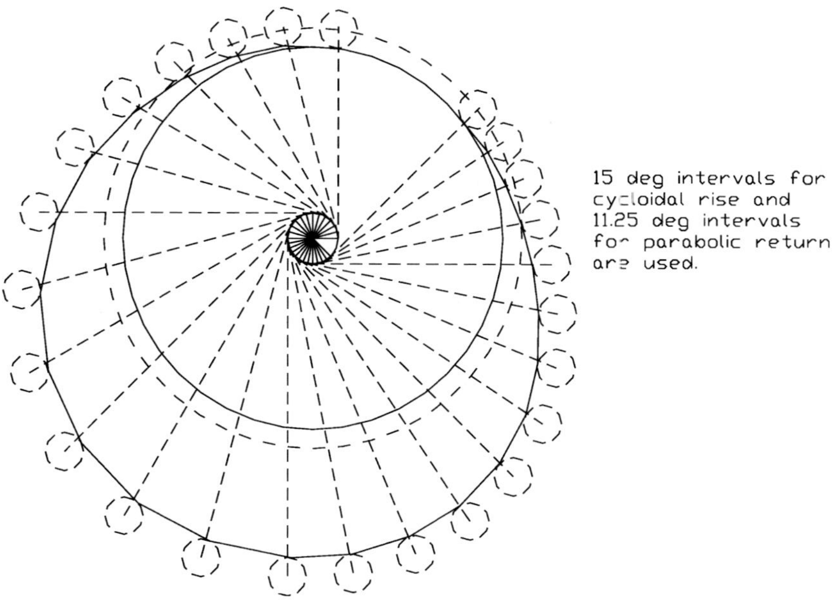

Fig. 5.38

θ	y	Trace point	
13	191.25	4.9306	13.2409
14	202.50	4.7222	13.0326
15	213.75	4.3750	12.6854
16	225.00	3.8889	12.1993
17	236.25	3.2634	11.5743
18	247.50	2.5000	10.8104

Parabolic motion return

$$\text{(Second half)} \quad y = 2L\left(\frac{\theta}{\beta}\right)^2$$

19	258.75	1.7361	10.0465
20	270.00	1.1111	9.4215
21	281.25	0.6250	8.9354
22	292.50	0.2778	8.5882
23	303.75	0.0694	8.3798
24	315.00	0.0000	8.3104

Solved Problem 5.3

Construct the profile of a disk cam with translating flat follower with the following data: Rise 3 cm with harmonic motion in 180° of cam rotation, return with parabolic motion in 150° of cam rotation and then dwell. The base circle diameter is 15 cm. Determine the width of the follower face allowing 0.75 cm clearance. The cam rotates in a counter-clockwise direction. Check the possibility for a cusp on the cam.

1. For harmonic motion $f(\theta) = 1.5(1 - \cos\theta)$ and $f'(\theta) = 1.5\sin\theta$. These values are tabulated in the table given below.
2. The width of the follower required in accordance with the harmonic rise motion above is $w = 2f'(\theta)_{\max} + 2 \times 0.75 = 4.5$ cm.
3. For parabolic displacement upto the inflection point in the return motion for the first 75° from 180° to 225°

$$f(\theta) = 3.0\left[1 - 2\left(1 - \frac{\theta}{\beta}\right)^2\right] \quad \text{and} \quad f'(\theta) = \frac{12}{\beta}\left(1 - \frac{\theta}{\beta}\right)$$

   These values are given in the table.
4. For parabolic displacement in the return motion $f(\theta) = 6.0(\theta/\beta)^2$ and $f'(\theta) = 12.0(\theta/\beta^2)$. These values are also tabulated. Figure 5.39 gives the cam profile.
5. For parabolic motion, at the point of inflection, $w = 2 \times 2.292 + 2 \times 0.75 = 6.084$ cm. Therefore the minimum width of the follower recommended is 6.1 cm.
6. In harmonic motion $f(\theta) = 1.5(1 - \cos\theta)$ and $f'(\theta) = 1.5\sin\theta$ and $R_c + f(\theta) + f'(\theta) = 9.0 > 0$, therefore there is no cusp here.
7. In parabolic motion

$$R_c + f(\theta) + f'(\theta) = 7.5 + 3.0\left(\frac{\theta}{\beta}\right)^2_{\min} + \frac{6 \times 180^2}{\pi^2\beta^2} = 8.3754 > 0$$

   therefore no cusp occurs in this motion either.

θ	$f(\theta)$	$f'(\theta)$	x	y
0	0.0000	0.0000	7.5000	0.0000
10	0.2280	0.2605	7.3633	1.5628
20	0.0946	0.5130	6.9572	3.0782
30	0.2010	0.7500	6.2942	4.5000
40	0.3509	0.9642	5.3944	5.7851
50	0.5358	1.1491	4.2851	6.8944
60	0.7500	1.2990	3.0000	7.7942
70	0.9870	1.4095	1.5782	8.4572
80	1.2395	1.4772	0.0628	8.8632
90	1.5000	1.5000	–1.5000	9.0000

θ	$f(\theta)$	$f'(\theta)$	x	y
100	1.7605	1.4772	−3.0628	8.8633
110	2.0130	1.4095	−4.5782	8.4572
120	2.2500	1.2990	−6.0000	7.7942
130	2.4642	1.1491	−7.2851	6.8944
140	2.6491	0.9642	−8.3944	5.7851
150	2.7990	0.7500	−9.2942	4.5000
160	2.9095	0.5130	−9.9572	0.0782
170	2.9772	0.2605	−10.3633	1.5628
180	3.0000	0.0000	−10.5000	0.0000
190	2.9733	−0.3056	−10.3673	−1.5177
200	2.8933	−0.6112	−9.9756	−2.9804
210	2.7600	−0.9167	−9.3438	−4.336
220	2.5733	−1.2223	−8.5023	−5.5387
230	2.3333	−1.5279	−7.4912	−6.5507
240	2.0400	−1.8335	−6.3579	−7.3451
250	1.6933	−2.1390	−5.1543	−7.9073
260	1.3067	−2.1390	−3.6358	−8.3015
270	0.9600	−1.8335	−1.8335	−8.4600
280	0.6667	−1.5279	−0.0866	−8.3079
290	0.4267	−1.2223	1.5625	−7.8667
300	0.2400	−0.9167	3.0761	−7.1614
310	0.1067	−0.6112	4.4213	−6.2200
320	0.0267	−0.3056	5.5693	−5.0721
330	0.0000	0.0000	6.4952	−3.7500
340	0.0000	0.0000	7.0477	−2.5652
350	0.0000	0.0000	7.3861	−1.3024
360	0.0000	0.0000	7.5000	0.0000

You can construct the displacement diagrams graphically and also plot the cam profile in accordance to the construction procedure given in Figure 5.26.

Solved Problem 5.4

The follower in the above problem is offset by 2 cm. Construct the cam profile.

Follow the construction from Figure 5.40. The cam profile itself is obtained by drawing a smooth curve tangential to all the flat faces of the follower.

Solved Problem 5.5

An oscillating flat follower driven by a disk cam is to rise through an arc of 25° with parabolic motion in 135° of cam rotation, dwell for 45° and return with cycloidal motion. Construct the cam surface and determine the dimensions of the flat follower. Cam rotation is clockwise. Choose the minimum size of the cam to be 4 cm and radius of oscillation to be 6 cm.

The lift is considered directly in degrees rather than a linear measure by converting the arc distance. At the starting position, full lines in Figure 5.41, the flat face is

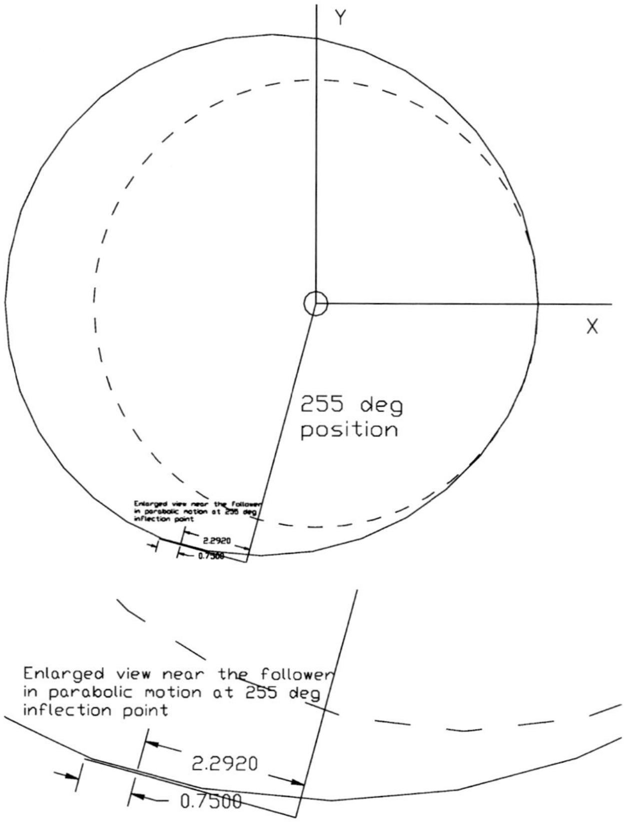

Fig. 5.39

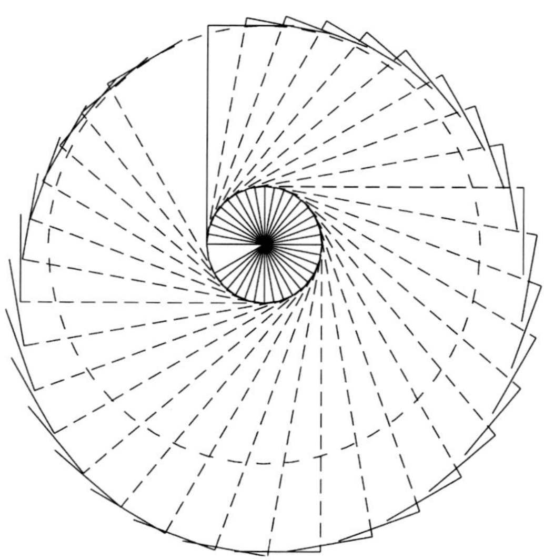

Fig. 5.40

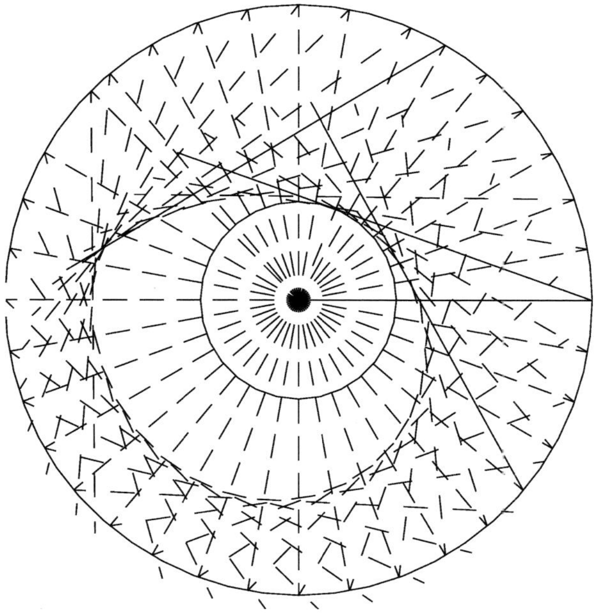

Fig. 5.41

at an angle given by $\sin^{-1}(2/6) = 19.4712°$. The angular displacement, rise, dwell and fall are tabulated below.

Parabolic motion		Cycloidal motion	
θ	ang. displ.	θ	ang. displ.
0	0.0000	180	25.0000
10	0.2743	190	24.9720
20	1.0974	200	24.7800
30	2.4691	210	24.2791
40	4.3896	220	23.3629
50	6.8587	230	21.9740
60	9.8765	240	20.1125
67.5	12.5000	250	17.8353
70	13.4088	260	15.2497
80	16.7010	270	12.5000
90	19.4444	280	9.7503
100	21.6392	290	7.1647
110	23.2853	300	4.8875
120	24.3827	310	3.0260
130	24.9314	320	1.6371
135	25.0000	330	0.7209
140	25.0000	340	0.2202
180	25.0000	350	0.0280
		360	0.0000

1. Draw the base circle of 2 cm radius.
2. Draw the follower center circle of 6 cm radius.
3. Locate the starting position shown in full lines.
4. Note that the angle made by the follower from the horizontal is 19.4712°.
5. Locate the follower center at the 10° position in a direction opposite to the cam rotation.
6. The angle made by the follower in this position is $19.4712 + 0.2743 = 19.7455°$.
7. Repeat for all positions given in the above table.
8. The cam surface is a smooth surface drawn tangential to all the follower surface lines.

Solved Problem 5.6

For an oscillating roller follower, construct the cam profile for the follower to have the same motion as in the above problem. The roller radius is 0.75 cm.

1. In the starting position the oscillating arm makes an angle with the horizontal given by
$$2\sin^{-1}\frac{2.75}{2\times 6} = 26.496°$$
2. At every interval, the angle made by the oscillating arm from the corresponding position of the starting point horizontal line, is given by the sum of 26.496° and the lift at the interval given in the table of the solved problem 5.5. e.g., in the first position this angle is $26.496 + 0.2743 = 26.7703°$.
3. Construct the locations of the oscillating arm 6 cm long at all the chosen intervals.
4. Draw circles of 7.5 mm radius at the tip points of these arms.
5. Then draw a smooth curve tangential to all these rollers, which gives the cam surface (see Figure 5.42 for the cam profile).

Solved Problem 5.7

A translating roller follower is to rise with acceleration, and rise further with a constant velocity of 4 m/s through a lift of 2 cm, rise yet again with deceleration and then finally fall through the total lift. A minimum dwell period of 45° is required before rising again. The total lift could be 3 to 4 cm. Design a suitable cam and follower to keep the inertia forces and jerks to a minimum and pressure angle within 45°. Check if there is any undercutting on the cam. The cam rotates in a clockwise direction at 3000 RPM.

Solution 1

1. Rise angle for the uniform velocity portion is

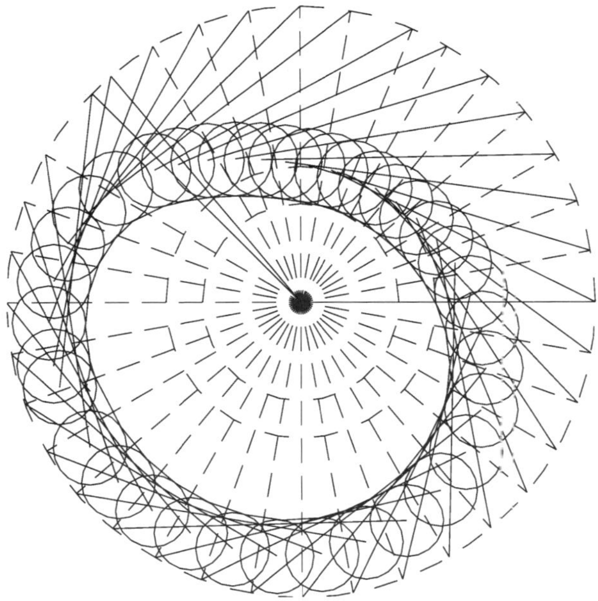

Fig. 5.42

$$\beta = \frac{L\omega}{v} = \frac{2 \times 3000 \times 360}{60 \times 400} = 90^\circ$$

2. To minimize the inertia forces and jerks, choose C-1 (curve 1) to connect with the uniform velocity portion (curve 2) and then H-2 (curve 3) to complete the rise.
3. The return motion can be chosen to be P-2 (curve 4) to couple with H-2.
4. Notice that all the kinematic conditions are matched to obtain continuity at the junction points (see Figures 5.13 and 5.14).
5. The dwell should be a minimum of 45°. Therefore choose a cam angle of 45° for C-1. The lift for this motion is given by (from equation 5.26)

$$v_{\theta=\beta_1} = 400 = \frac{2L\omega}{\beta} \quad \text{or} \quad L_1 = 0.5 \text{ cm}$$

6. For H-2 motion,

$$v_{\theta=0} = 400 = \frac{L\omega\pi}{2\beta} \quad \text{for} \quad L_3 = 0.5 \text{ cm}$$

This gives

$$\beta_3 = \frac{\pi^2}{16} = 35.34^\circ$$

(see Figure 5.13 for the displacement and differentiate this to obtain the velocity).

7. The end acceleration of H-2 motion is

$$y''_{\theta=\beta} = -\frac{L\pi^2}{4\beta^2}\sin\frac{\pi\theta}{2\beta}$$

giving $\alpha = -32/\pi^2$.

8. Hence for P-2 motion (Figure 5.15)

$$-5.2683\frac{L}{\beta^2} = -\frac{32}{\pi^2}$$

giving $\beta_4 = 126.5°$.

9. Choice of base circle: From the pressure angle relation (5.45), for uniform motion with $R = R_0 + L(\theta/\beta)$, $\tan\alpha = 1$ gives

$$R_0 = \frac{L}{\beta}(1-\theta) = \frac{4}{\pi} - 2\frac{\theta}{\beta}$$

The base circle radius thus varies from 1.273 to −0.7268 with the angle of rotation from zero to 90°, for a pressure angle of 45°. Therefore choose the base circle radius to be at least 1.273 cm.

10. Choosing the minimum radius of the pitch surface of the cam to be 3.25 cm, we find that the pressure angle is less than 45° for all the other three motions.
11. Choice of the roller: This is chosen so as to avoid any undercutting. The straight line motion is

$$R = R_0 + f(\theta) = 3.25 + 2\frac{\theta}{\beta}$$

12. From equation (5.42)

$$\rho = \frac{\left[\left(3.25 + 2\frac{\theta}{\beta}\right)^2 + \frac{16}{\pi^2}\right]^{1.5}}{\left[\left(3.25 + 2\frac{\theta}{\beta}\right)^2 + \frac{32}{\pi^2}\right]}$$

13. The radius of curvature given above varies from 3.08 to 5.12 cm, as the cam angle varies from zero to 90°. Hence the roller radius should be less than 3.08 cm.
14. For the polynomial return motion,

$$\frac{L}{R_0} = \frac{3}{3.25} = 0.923, \quad \beta = 126.5°$$

The condition for no undercutting from Figure 5.23 is $\rho_{min}/R_0 = 1.2$, hence $\rho_{min} = 3.9$.

15. Choosing the roller radius to be 0.75 cm, we find that there is no undercutting for all the other motions.
16. The following table gives values in cm and degrees for the pitch surface of the cam.

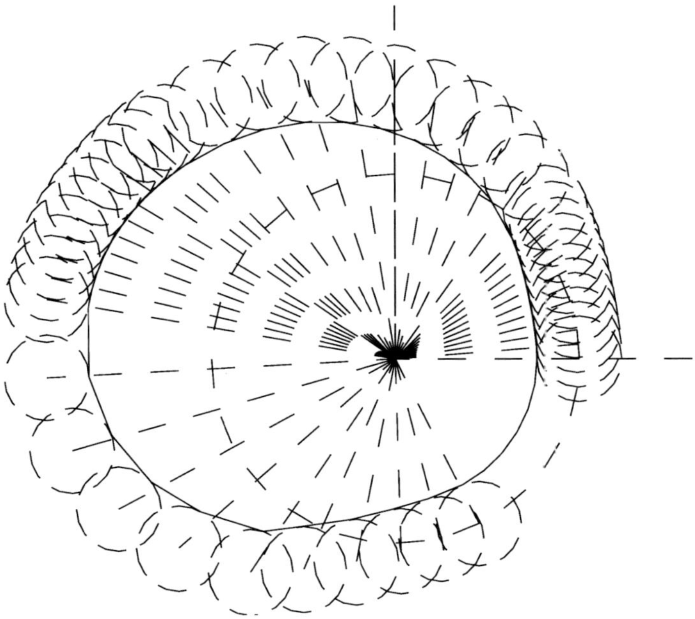

Fig. 5.43

C-1		H-2	
3.25082	4.5	5.82822	138.534
3.25645	9.0	5.90451	142.068
3.27124	13.5	5.97700	145.602
3.29863	18.0	6.04389	149.136
3.34085	22.5	6.10355	152.670
3.39863	27.0	6.15451	156.204
3.47124	31.5	6.19550	159.738
3.55645	36.0	6.22553	163.272
3.65082	40.5	6.24384	166.806
3.75000	45.0	6.25000	170.340
Uniform motion		P-2	
3.95	54.0	6.171	182.99
4.15	63.0	5.937	195.64
4.35	72.0	5.562	208.29
4.55	81.0	5.081	220.94
4.75	90.0	4.552	235.59
4.95	99.0	4.049	246.24
5.15	108.0	3.642	258.89
5.35	117.0	3.381	271.54
5.55	126.0	3.268	284.19
5.75	135.0	3.250	296.84

The final cam surface is obtained by drawing a tangent to all the roller positions as given in Figure 5.43.

Solution 2.
There can be several solutions for such problems like this. One more alternative is given here.

1. Choose an eighth-power polynomial to connect the uniform velocity portion on either side of the rise and fall portions. Here the dwell should be minimum of 45°. Therefore we can choose a total of 150° for rise and fall portions with a dwell of 60°. Each of the four eighth-power polynomials will thus take 30°. The linear motion is sandwiched between two eighth-power polynomials in both the rise and fall portions of the displacement diagram.
2. Choice of base circle: The condition for the uniform motion is already given in item 9 of the Solution 1 discussed above.
3. For the eighth-power polynomial portion with the cam angle 30° and pressure angle 45°, L/R ratio is 0.4. For $L = 1$, the minimum base circle radius required is 2.5 cm. Let us choose this value to be 3.25 cm.
4. Choice of the roller: This is chosen so as to avoid any undercutting. The condition for uniform motion is already discussed in item 11 of solution 1. For the polynomial motion, the condition for no undercutting is, $L/R = 1/3.25$ and cam angle 30°, therefore, $\rho_{\text{min}}/R_0 = 0.25$ hence $\rho_{\text{min}} = 0.8125$. The radius of the roller should be less than this and it is chosen as 0.75 cm.
5. Displacement values: From the eighth-power polynomial equation in Figure 5.15, the displacement values for the entire 360° motion are calculated and given in the table below. The cam surface is drawn in Figure 5.44.

θ		S
0	360	0.0
6	294	0.04367
12	288	0.2663
18	282	0.6104
24	276	0.8950
30	270	1.0000
40	260	1.2222
50	250	1.4444
60	240	1.6666
70	230	1.8888
80	220	2.1111
90	210	2.3333
100	200	2.5555
110	190	2.7777
120	180	3.0000
126	174	3.04367
132	168	3.2663
138	162	3.6104
144	156	3.8950
150	150	4.0000

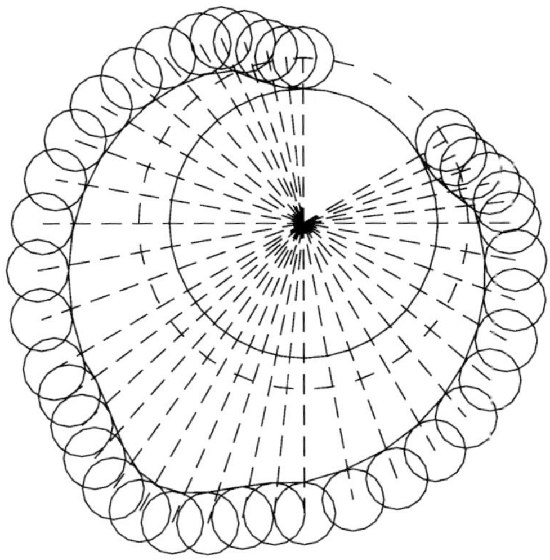

Fig. 5.44

Solved Problem 5.8

A translating flat follower is required to rise through a lift of 4 cm during 120° of a high speed cam rotating at 3600 RPM in the counter-clockwise direction and then fall through 120°, to be followed by a dwell. Recommend a suitable cam and follower with a proper displacement function. Make a layout of the cam profile by graphical methods and check it with the parametric equations of the cam surface.

An eighth-power polynomial is chosen to minimize the inertia forces and jerks. The displacement and its derivative along with the x and y coordinates are given below.

θ	$f(\theta)$	$f'(\theta)$	x	y
0	0.0000	0.0000	4.0000	0.0000
11	0.0183	0.2814	3.8908	1.0430
21	0.1199	0.9291	3.5134	2.3438
31	0.3527	1.7491	2.8301	3.7410
41	0.7282	2.5322	1.9071	5.0131
51	1.2245	3.1094	0.8714	6.0170
61	1.7955	3.3787	−0.1454	6.7069
71	2.3841	3.3101	−1.0513	7.1139
81	2.9331	2.9355	−1.8148	7.3070

θ	$f(\theta)$	$f'(\theta)$	x	y
91	3.3952	2.3275	–2.4562	7.3535
101	3.7372	1.5742	–3.0216	7.2947
111	3.9410	0.7537	–3.5495	7.1435
120	4.0000	0.0000	–4.0000	6.9282
131	3.9115	–0.9203	–4.4959	6.5746
141	3.6794	–1.7324	–4.8778	6.1791
151	3.3115	–2.4635	–5.2005	5.6993
161	2.8289	–3.0318	–5.4698	5.0899
171	2.2678	–3.3500	–5.6666	4.2893
181	1.6780	–3.3519	–5.7356	3.2523
191	1.1175	–3.0164	–5.5990	1.9845
201	0.6423	–2.3875	–5.1896	0.5653
211	2.9453	–1.5817	–4.4957	–0.8561
221	0.0901	–0.7788	–3.5978	–2.0956
231	0.0101	–0.0191	–2.6720	–2.9963
240	0.0000	0.0000	–2.0000	–3.4641

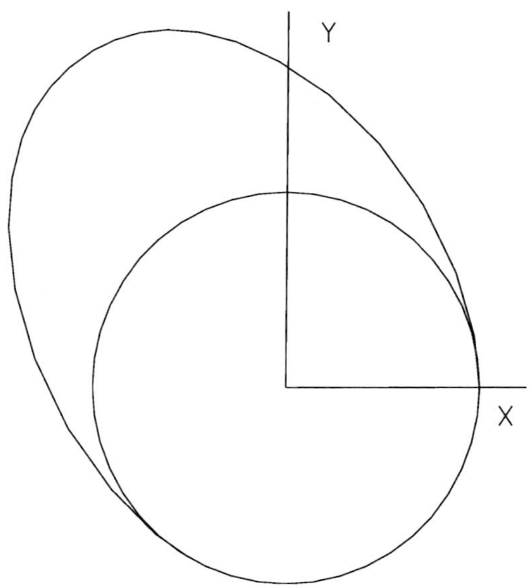

Fig. 5.45

Figure 5.45 shows the plot of the above coordinates which gives the cam surface. Figure 5.46 is the construction by a conventional graphical method and the cam surface is identically the same as in Figure 5.45.

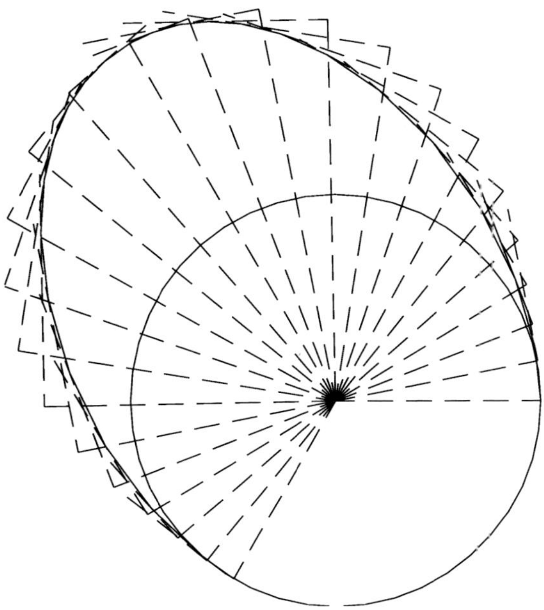

Fig. 5.46

Solved Problem 5.9

A translating roller follower of 1 cm radius is driven by an eccentric circle cam of radius 3.5 cm and eccentricity 1.2 cm, rotating at 200 RPM. Determine the displacement, velocity and acceleration of the cam for one full rotation. Also, determine the maximum pressure angle.

1. From equation (5.57),

$$y = E\,(1 - \cos\theta) - l + \sqrt{l^2 - E^2 \sin^2\theta}$$

2. $l = 3.5 + 1.0 = 4.5$ and $l/E = 3.75$.
3. In view of item 2 above, we can make an approximation for the square root term in the above and use just one term,

$$x = E - l - E\cos\theta + l\left(1 - \frac{E^2}{2l^2}\sin^2\theta\right)$$

4. $v = \omega\left(E\sin\theta - \dfrac{E^2}{2l}\sin 2\theta\right)$
5. $a = \omega^2\left(E\cos\theta - \dfrac{E^2}{l}\cos 2\theta\right)$
6. From (5.45),

$$\alpha = \tan^{-1} 1.2 \frac{\sin\theta - \frac{1}{7.5}\frac{\sin 2\theta}{\sqrt{1-\frac{\sin^2\theta}{14.0625}}}}{-1.0 - 1.2\cos\theta + 4.5\sqrt{1-\frac{\sin^2\theta}{14.0625}}}$$

You can tabulate the above values as a function of θ, the maximum vlue of pressure angle, 21.691°, occurs at θ equal to 85.5 and 274.5°.

Solved Problem 5.10

A translating flat follower is to be driven by an eccentric circle cam of the above problem. Determine the displacement, velocity and acceleration of the follower. Also determine the width of the follower with a clearance of 0.75 cm on either side.

$$y = 1.2\,(1 - \cos\theta)\,, \quad y' = 1.2\sin\theta \quad \text{and} \quad y'' = 1.2\cos\theta$$

$$v = 25.1328\sin\theta \text{ cm/s} \quad \text{and} \quad a = 526.38\cos\theta \text{ cm/s}^2$$

$$w = 2y'(\theta) + 2\times 0.75, \quad w_{\min} = 2.4 + 1.5 = 3.9 \text{ cm}$$

Solved Problem 5.11

A triple curve cam rotating at 900 RPM is to have a total lift of 1.25 cm for a cam rotation of 75°. The maximum distance of cam profile from its center is 4.5 cm. The follower roller radius is 1 cm and the nose radius 1.5 cm. Plot the profile of the cam. Determine the displacement, velocity and acceleration expressions for one full cycle. Also determine the pressure angle for one full revolution and determine its maximum value and the location where it occurs. Check whether there is a possibility of undercutting.

1. $L + R_c = 4.5 = E_2 + \rho_n$.
2. $R_c = 4.5 - 1.25 = 3.25$ cm and $E_2 = 4.5 - 1.5 = 3.0$ cm.
3. $E_1 = E_2 = E_3 = 3.0$ cm.
4. $\gamma = 180 - 75 = 105°$.
5. Figure 5.47 gives the cam and follower for $\theta = 150°$, then equation (5.67) gives
6.
$$y = E_1 - l_1 - E_1\cos(\theta - 105) + \sqrt{l_1^2 - E_1^2\sin^2(\theta - 105)}$$
$$= 4.25 - 3\cos 45 + \sqrt{7.25^2 - 9\sin^2 45}$$
$$= 0.56139$$
7. $R = -3\cos(\theta - 105) + \sqrt{7.25^2 - 9\sin^2(\theta - 105)}$.
8. $\dfrac{dR}{d\theta} = 3\sin(\theta - 105) - \dfrac{4.5\sin 2(\theta - 105)}{\sqrt{7.25^2 - 9\sin^2(\theta - 105)}}$.

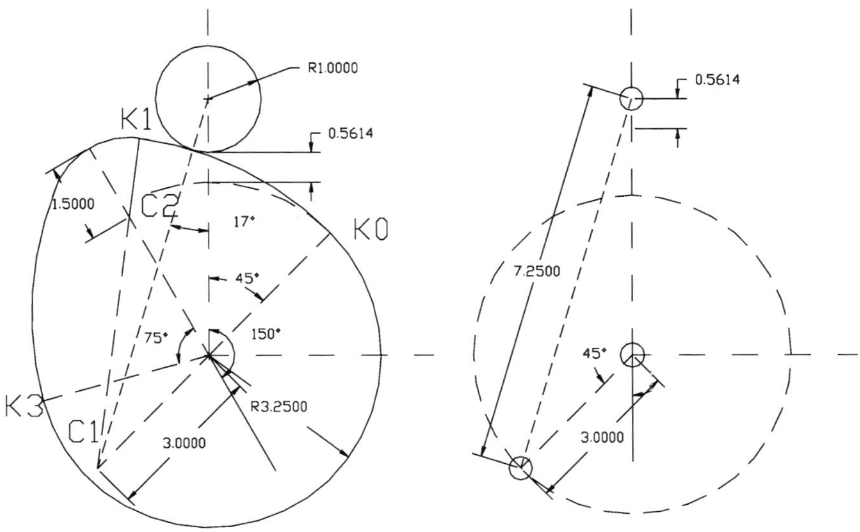

Fig. 5.47

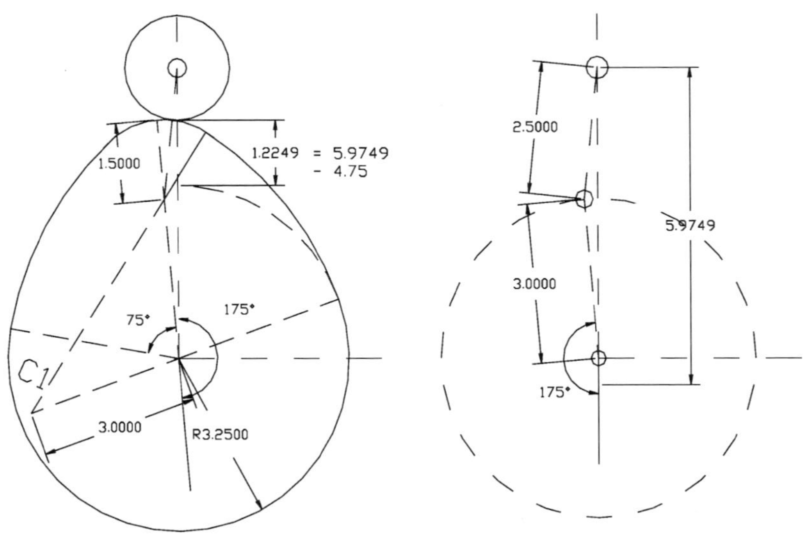

Fig. 5.48

9. $\alpha = \tan^{-1} \dfrac{1}{R}\dfrac{dR}{d\theta} = \tan^{-1} 0.30599 = 17.01°$.
The pressure angle above can also be determined from the equivalent mechanism as $7.25/\sin 135 = 3/\sin\alpha$, therefore

$$\alpha = \sin^{-1} \frac{3 \times 0.7071}{7.25} = 17.01^\circ$$

10. Figure 5.48 gives the cam and follower for $\theta = 150^\circ$, then equation (5.64) gives $X = 2\rho_N - 2R_c - L = -4.75$ and
11.
$$\begin{aligned} y &= E_2 - l_2 - E_2 \cos\theta + \sqrt{l_2^2 - E_2^2 \sin^2\theta} + X \\ &= -4.25 - 3\cos 175 + \sqrt{6.25 - 9\sin^2 175} \\ &= 1.22487 \end{aligned}$$

12. $R = -3\cos\theta + \sqrt{6.25 - 9\sin^2\theta}$.
13. $\dfrac{dR}{d\theta} = 3\sin\theta - \dfrac{4.5\sin 2\theta}{\sqrt{6.25 - 9\sin^2\theta}}$.
14. $\alpha = \tan^{-1} \dfrac{1}{R}\dfrac{dR}{d\theta} = \tan^{-1} 0.10516 = 6.0^\circ$.
15. Also
$$\alpha = \sin^{-1} \frac{3 \times \sin 5}{2.5} = 6.0^\circ$$
16. The minimum radius of curvature of the cam is 1.5 cm, which is more than the radius of the roller follower. Hence, there will be no undercutting or pointing.

Solved Problem 5.12

The triple curve cam of the above problem is to drive a translating flat follower. Determine the maximum velocity and acceleration. Also determine a suitable width for the follower. Show that no cusps occur for the cam.

1. For the first curve (Figure 5.49) $y = 3 - 3\cos(130 - 105) = 0.2811$, see equations (5.58) and (5.59).
2.
$$\begin{aligned} y' &= 3\sin(\theta - 105) \\ y'' &= 3\cos(\theta - 105) \\ R_c + y + y'' &= 6.25 > 0 \end{aligned}$$

3. For the second curve (Figure 5.50), $y = -1.75 - 3\cos 175 = 1.2386$, see equations (5.60) and (5.65) and the value of X from the previous problem. (Note a small error in the third decimal place in the graphical construction, arising from drawing the flat face to touch the cam surface.)
4.
$$\begin{aligned} y' &= 3\sin\theta \\ y'' &= 3\cos\theta \\ R_c + y + y'' &= 1.5 > 0 \end{aligned}$$

5. Hence there is no cusp on the cam surface.

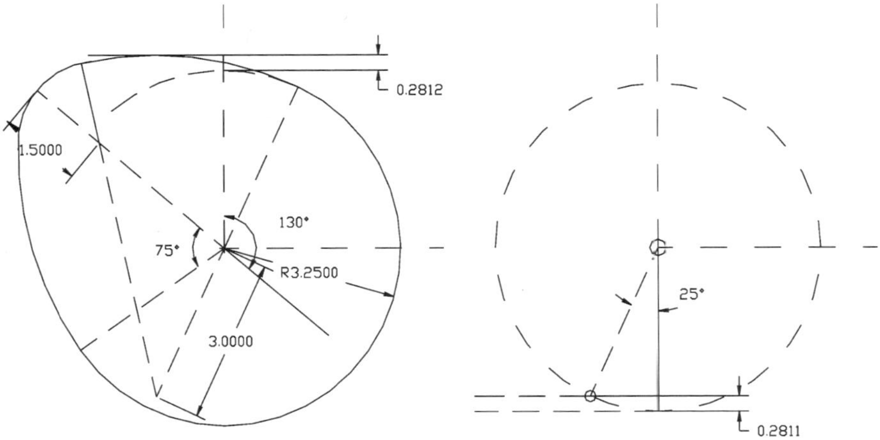

Fig. 5.49

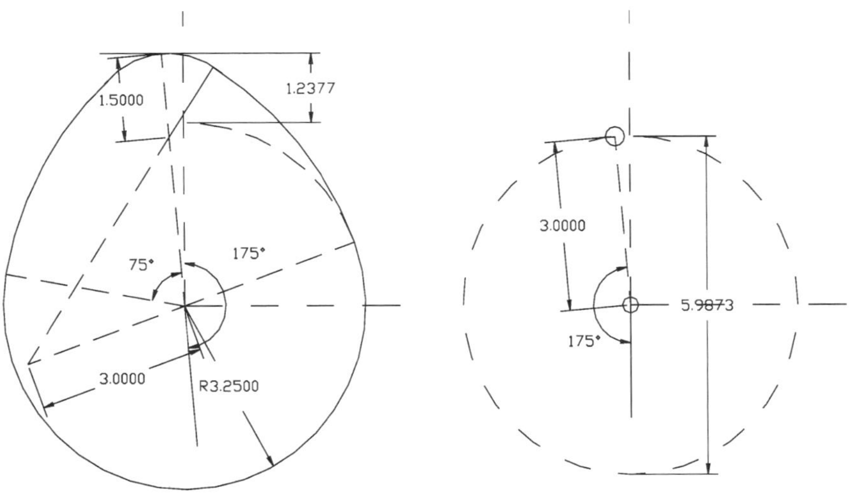

Fig. 5.50

Solved Problem 5.13

A tangent cam is to drive a roller follower through a total lift of 1.25 cm for a cam rotation of 75°. The cam speed is 600 RPM. The distance between cam center and follower center at full lift is 4.5 cm and the roller is 2 cm in diameter. Find the cam proportions and determine the displacement for 10° and 65° of rotation of the cam in the lift portion.

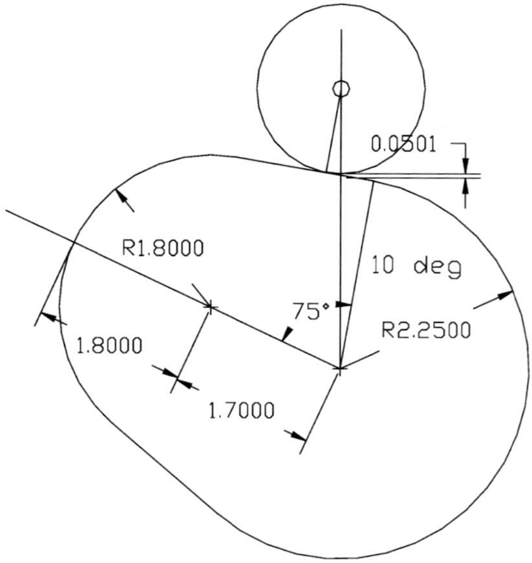

Fig. 5.51

1. $4.5 = R_c + 1.25 + R_r$, therefore $R_c = 2.25$ cm.
2. For 10° of cam rotation in the lift portion, Figure 5.51, the displacement is given by $y = 3.25(\sec\theta + 1)$, see equation (5.73).
3. $R = R_0 + f(\theta) = 3.25\sec\theta$.
4. $\dfrac{dR}{d\theta} = 3.25\sec\theta\tan\theta$.
5. $\alpha = \tan^{-1}\tan\theta = \theta$.
6. Refer Figure 5.52 for cam rotation 65° into the lift portion, i.e., $\theta = 170°$, the displacement
7. $y = E_2(1-\cos\theta) - l_2 + \sqrt{l_2^2 - E_2^2\sin^2\theta} + 2\rho_N - 2R_c - L$, see equation (5.70).
   $y = 1.7(1-\cos\theta) - 2.8 + \sqrt{2.8^2 - 1.7^2\sin^2\theta} + 2\times 2.25 - 1.25 = 1.2086$.
8. $R = R_0 + f(\theta) = 4.4586$.
9. $\dfrac{dR}{d\theta} = 1.7^2\sin\theta - \dfrac{1.7\sin 2\theta}{2\sqrt{2.8^2 - 1.7^2\sin\theta}} = 0.47276$.
10. $\alpha = \tan^{-1}\dfrac{0.47276}{4.4586} = 6.05°$.
11. Since the minimum radius of curvature of the cam is 1.8 cm which is more than the roller radius, no undercutting occurs.

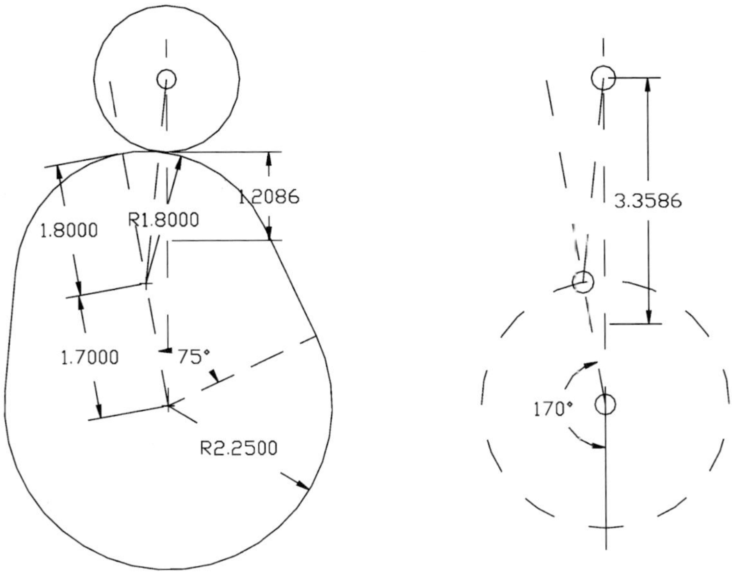

Fig. 5.52

5.9 Additional Problems

1. A push rod operated by a cam, is to rise and fall with simple harmonic motion along an inclined straight path. The least radius of the cam is 5 cm and the push rod is fitted at its lower end with a roller of 3.75 cm diameter. When in its lowest position, the roller center is vertically above the cam axis. The maximum displacement of the roller is 5 cm in a direction 30° to the right of the vertical. The cam rotates at 100 RPM in a clockwise direction. The time of lift is 0.15 sec and the time of fall 0.10 sec with a period of rest of 0.05 sec at the upper position. Draw the cam profile.
2. In the above problem, obtain the cam profile if the roller follower is to be replaced by a flat faced follower.
3. A cam rotating on a horizontal axis operates a vertical rod which is fitted with a roller 4 cm diameter at its lowest end. The axis of the rod does not pass through the cam axis, but is offset, 2 cm. The rod rises during 60° rotation of the cam, falls during the following 60°, and remains at rest for the following 240°. The rod moves with parabolic motion during both strokes. Taking the least radius of the cam to be 5.5 cm, draw the profile of the cam.
4. In the above problem, obtain the cam profile if the roller follower is to be replaced by a flat faced follower.
5. An oscillating roller follower is required to operate through an angle of 30° with cycloidal motion in a cam angle of rotation 120° for rising motion and return without a dwell through 150° of cam rotation. The minimum size of the cam is

5 cm. The center of the oscillating follower is located at 12 cm in the horizontal direction to the right of the cam center and 8 cm vertically upwards. The radius of the lever is 11 cm at the tip of which the roller of radius 1.5 cm is mounted. Draw the cam profile.

6. In the above problem, the lever should be at the same location at the starting point of lift, however, it should carry at its end a flat surface instead of the roller. Draw the cam surface to obtain the same motion for the lever and also specify the geometry of the flat face follower.
7. A follower is required to move at a constant velocity of 5 cm/sec for one second. This motion can be preceded and followed by other motions so that the cam action is smooth without any jerks and returned to the starting point. The total cycle time is 2.75 seconds. Design a cam for a follower radius of 1.25 cm and a maximum pressure angle 30°. Make sure that there is no undercutting.
8. In the above problem, replace the roller follower by a flat follower and design a suitable cam. The possibility of any cusp should be avoided.
9. A circular cam plate of 7.5 cm diameter is used to lift a translating roller follower of diameter 3.75 cm. The distance between the axis of rotation and the center of the cam plate is 2.5 cm. The cam rotates at 60 RPM counter-clockwise. Determine the velocity of the follower when the cam center is at 45° from the horizontal line passing through the cam axis of rotation.
10. A cam with convex flanks, operating a flat follower whose lift is 3 cm, has a base radius of 6 cm and a nose radius of 1.6 cm. The cam is symmetrical about a line drawn through the center of curvature of the nose and the center of the camshaft. If the total angle of cam action is 120°, find the radius of the convex flanks. Determine the maximum velocity and the maximum acceleration and retardation when the cam shaft speed is 500 RPM.
11. Solve the above problem if the follower is a translating roller follower fitted with roller of radius 3 cm.
12. A valve is operated by a cam which has a base circle diameter 7 cm and a lift of 2.5 cm. The cam has tangent flanks and a circular nose and the total angle of action is 120°. The follower which has a roller of 3 cm diameter moves along a straight line passing through the cam axis. Find the maximum acceleration when the cam rotates at 1000 RPM.
13. A symmetrical cam has a base circle of 6 cm radius, arc of action 110°, and straight flanks with the tip of a circular arc. The line of action of the follower passes through the center line of the camshaft. The follower which has a 4 cm diameter roller has a lift of 2.6 cm. Calculate the velocity and acceleration of the follower when moving outwards and contact is just reaching the end of the straight flank, when the cam is rotating at 500 RPM.

Chapter 6
Spur Gears

Gear

Wheel with teeth on its surface designed to mesh with another gear or rack.

6.1 Classification of Gears

Cylindrical Gear

Gear with teeth formed on a cylindrical surface.

Spur Gear

Cylindrical gear with external teeth parallel to the axis of shaft, see Figure 6.1. The line of contact between the mating teeth is always parallel to the shaft axis.

Internal Spur Gear [Annulus]

Cylindrical gear with internal teeth.

Gear Sector [Segment]

Segment of a spur gear or annulus.

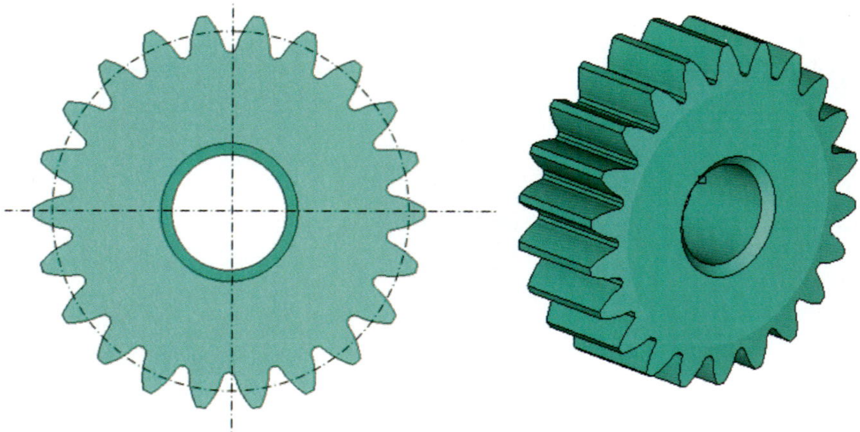

Fig. 6.1 Spur gear

Pinion

(1) The smaller of a pair of meshing cylindrical gears. (2) Cylindrical gear meshed with a rack.

Rack

Segment of a cylindrical gear of infinite radius.

Spur Rack and Pinion

Rack whose motion is perpendicular to the axis of pinion.

Helical Gear

Gear with teeth wrapped helically on a cylindrical surface, see Figure 6.2a.

Herring-Bone Gear [Double-Helical]

Gears comprising two integral helical gears, the helices of the gears being of opposite hand, see Figure 6.2b.

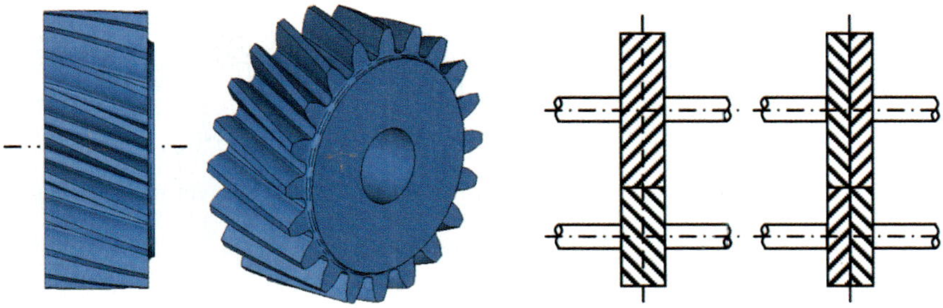

Fig. 6.2 (a) Helical gear. (b) Herringbone gear

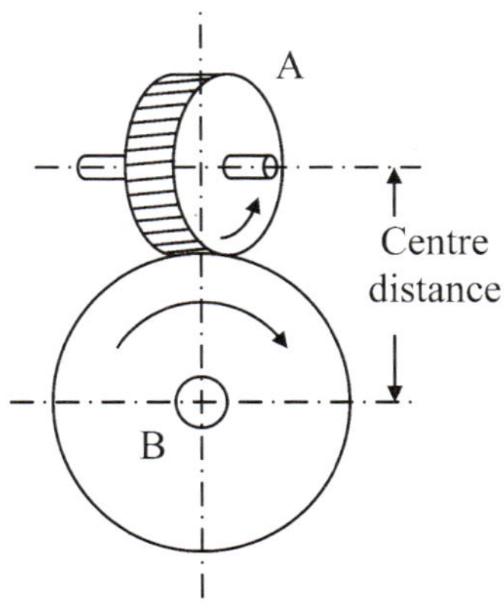

Fig. 6.3 Spiral gears

Spiral [Crossed-Helical] Gear

Helical gear used in transmission of motion between two skew shafts, see Figure 6.3.

Spiral Rack and Pinion

Rack whose motion is not perpendicular to the axis of pinion.

Bevel [Conical] Gear

Gear with teeth formed on a conical surface.

External Bevel Gear

Bevel gear with pitch angle of the cone is less than 90°, see Figure 6.4a.

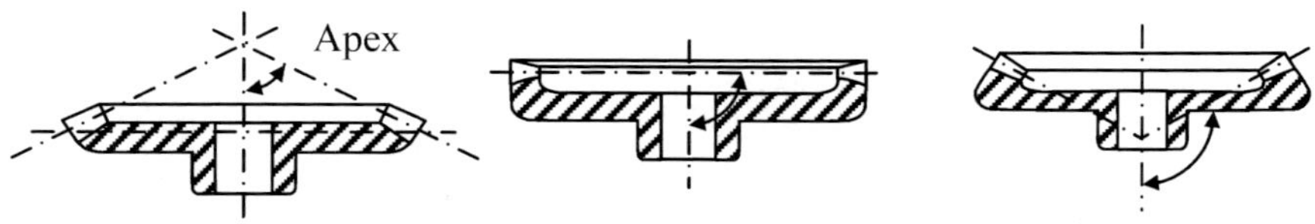

(a) External Bevel Gear (b) Crown Wheel c) Internal Bevel Gear

Fig. 6.4 Bevel gears

Crown Wheel

Bevel gear with pitch angle equal to 90°, see Figure 6.4b.

Internal Bevel Gear

Bevel gear with pitch angle less than 90° , see Figure 6.4c.

Straight Bevel Gear

Bevel gear whose teeth are radial to the point of intersection of the shaft axes.

Mitre Gear

Bevel gears with gear ratio 1:1 connecting two perpendicular shafts.

Spiral Bevel Gear

Spiral bevel gear has the same relationship with straight bevel gear, as helical gears have the relation with spur gears.

Hypoid Gear

Spiral bevel gear pair with offset between the gear axes.

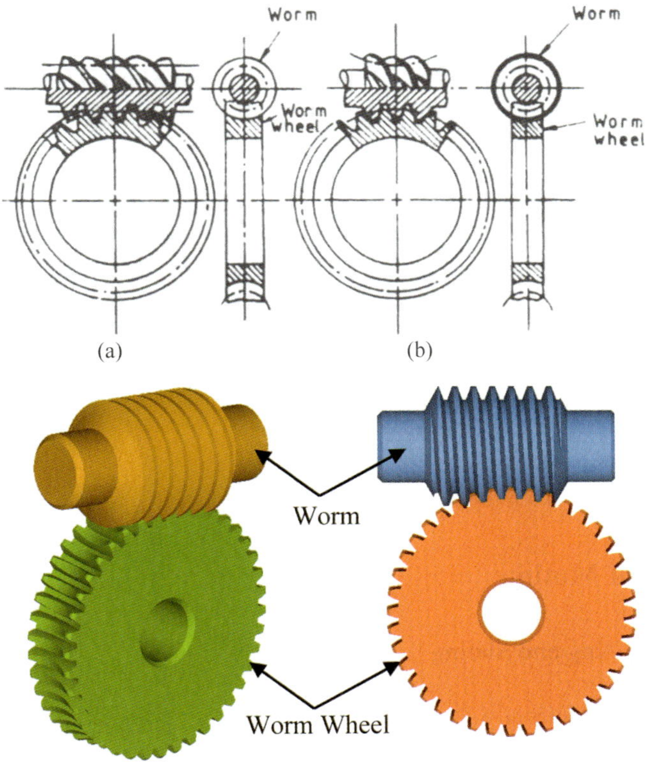

Fig. 6.5 Worm gears. (a) Straight worm. (b) Enveloping worm

Worm Gear

Gear with one or more teeth wrapped helically on a cylinder, the pitch of the helix being less than the diameter of the gear, see Figure 6.5a. It is also commonly known as worm or straight worm.

Enveloping Worm

An enveloping worm, see Figure 6.5b.

Worm Wheel

Gear that mates with a worm gear.

Idler

Gear intermediate between a driving and a driven gear, which affects the sense of direction of the latter but does not affect the velocity ratio.

Planetary [Planet] Gear

Gear that rotates on an axle whose own axis is constrained to rotate about another axis.

Sun Wheel

Gear with a mating planet.

6.2 Types of Motion

Combined Rolling and Sliding

Motion transmitted from a driver to a driven member with direct contact where rolling and sliding both occurs simultaneously.

Figure 6.6a illustrates this motion. P is the direct contact point at the instant shown. NN is the common normal and the normal component of velocity P_n of the common point in both the bodies is the same. The tangential components, P_2t_2 and P_3t_3 however, could be different as illustrated.

Pure Rolling

No single point on one of the members makes contact with two successive points on the second member. The two tangential components, P_2t_2 and P_3t_3 are same and there is no relative sliding (see Figure 6.6b). The point of contact between the two bodies lies on the line of centers.

Pure Sliding

Here, a single point belonging to one of the members makes contact with all the successive points of the second member. One of the members is at rest and there is no component of velocity in the normal direction (see Figure 6.6c).

In the general case, the velocity of the driving member 2 rotating at ω_2 rad/s is

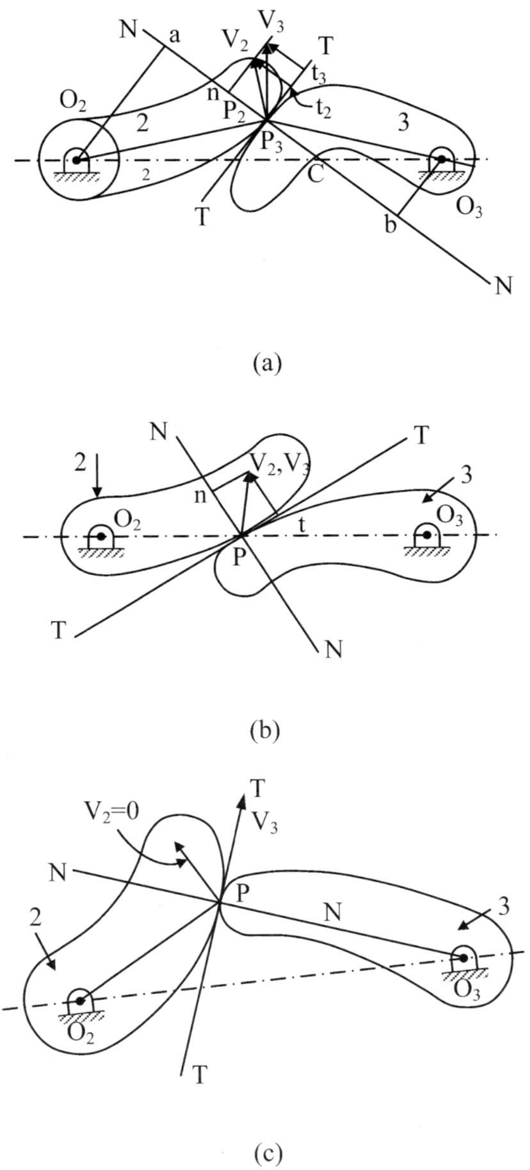

Fig. 6.6 (a) Combined rolling and sliding. (b) Pure rolling. (c) Pure sliding

$$P_2V_2 = O_2P_2 \times \omega_2 \tag{6.1}$$

Because the normal component should be the same for both bodies (otherwise there will be digging), we have

$$\omega_3 = \frac{P_3V_3}{O_3P_3} \tag{6.2}$$

and

$$\frac{\omega_3}{\omega_2} = \frac{P_3V_3}{O_3P_3}\frac{O_2P_2}{P_2V_2} \tag{6.3}$$

Follow the construction shown in the figure and confirm that O_2P_2a and nP_2V_2, and O_3bP_3 and P_3V_3n are two sets of similar triangles. Hence

$$\frac{\omega_3}{\omega_2} = \frac{P_3n}{O_3b}\frac{O_2a}{P_2n} = \frac{O_2a}{O_3b} \tag{6.4}$$

Further the triangles O_2aC and O_3bC are also similar, therefore

$$\frac{\omega_3}{\omega_2} = \frac{O_2C}{O_3C} \tag{6.5}$$

Angular Velocity Ratio

Therefore, the ratio of angular velocities for a pair of rigid bodies transmitting motion with direct contact is equal to the inverse ratio of segments into which the line of centers is divided by a common normal through the point of contact.

6.3 Nomenclature

Refer to Figure 6.7.

Pitch Circle

This is a right section of an imaginary cylinder (pitch cylinder), that the toothed gear may be considered for replacement.

Pitch Diameter, D

Diameter of pitch cylinder; also commonly called diameter of gear.

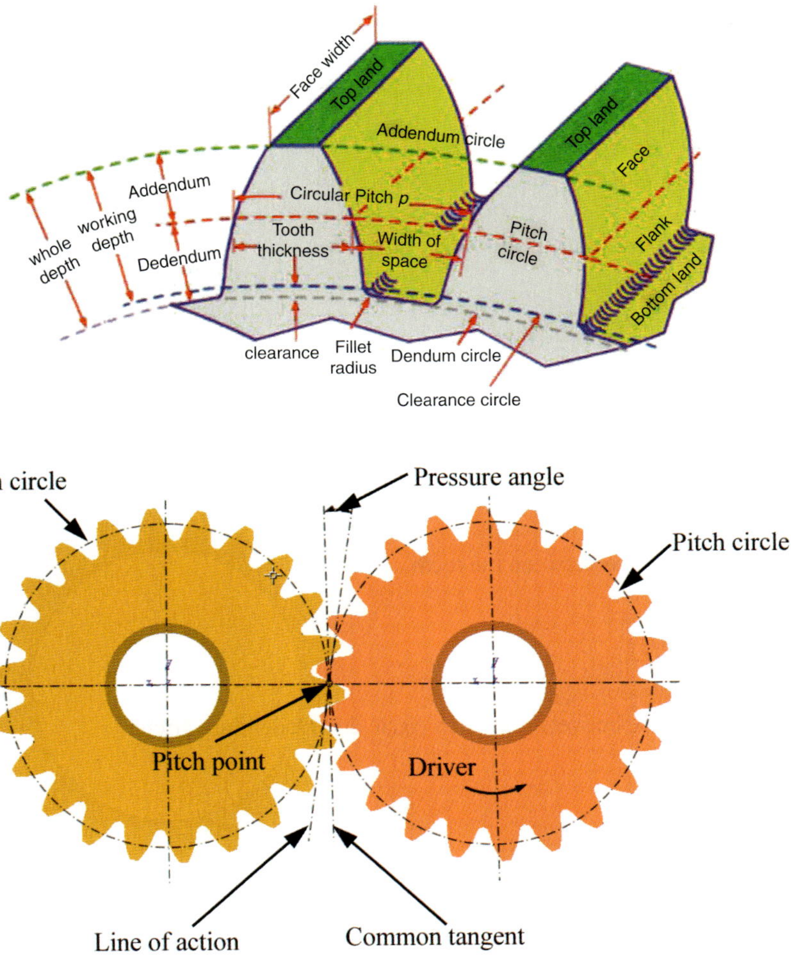

(a) (b)

Fig. 6.7 (a) Gear geeth representation. (b) Pair of mating gears

Pitch Point

The point of tangency of the pitch circles of a pair of mating gear wheels.

Common Tangent

Tangent to the pitch circles at the pitch point.

Common Normal

Normal to the mating profiles at the point of contact.

Addendum Circle

Circle that passes through all the tooth ends.

Addendum

Radial distance between pitch circle and addendum circles.

Dedendum [Root] Circle

Bounds the space between the teeth.

Dedendum

Radial distance between the pitch circle and Dedendum circle.

Clearance

The difference between Dedendum of one gear and addendum of the mating gear.

Top Land

Top surface of the tooth.

Bottom Land

Bottom surface of the tooth.

Face

Part of the tooth between the pitch cylinder and addendum cylinder.

Face Width

Width of the face along the tooth.

Flank

Part of the tooth between the pitch cylinder and Dedendum cylinder.

Tooth Thickness

Thickness of the tooth measured along the arc of the pitch circle.

Tooth Space

Circular distance between two successive teeth measured along the pitch circle.

Backlash

Difference between the tooth space of one gear and tooth thickness of the mating gear.

Circular Pitch, p_c

Sum of tooth thickness and tooth space, measured along a pitch circle.

Diametral Pitch, p_d

Number of teeth of a gear per unit pitch diameter.

Module, *m*

Inverse of diametral pitch.

The diametral pitch (module) should be the same for two mating gears (as well as circular pitch). If the number of teeth on the gear is T, then

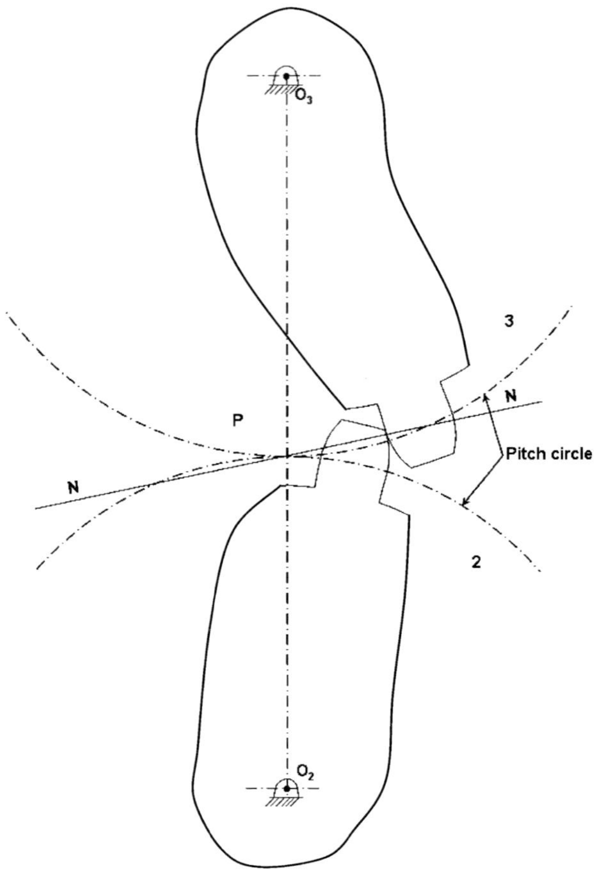

Fig. 6.8 Gear tooth action

$$p_c = \frac{\pi D}{T}$$

$$p_d = \frac{T}{D}$$

$$p_c \times p_d = \pi \tag{6.6}$$

6.4 Law of Gear Tooth Action

The common normal must always pass through the pitch point.

The above condition is shown in Figure 6.8. Equation (6.5) gives the speed ratio as $\omega_3/\omega_2 = O_2P/O_3P$. This ratio will remain constant as long as the common normal in all positions of mating passes through the pitch point.

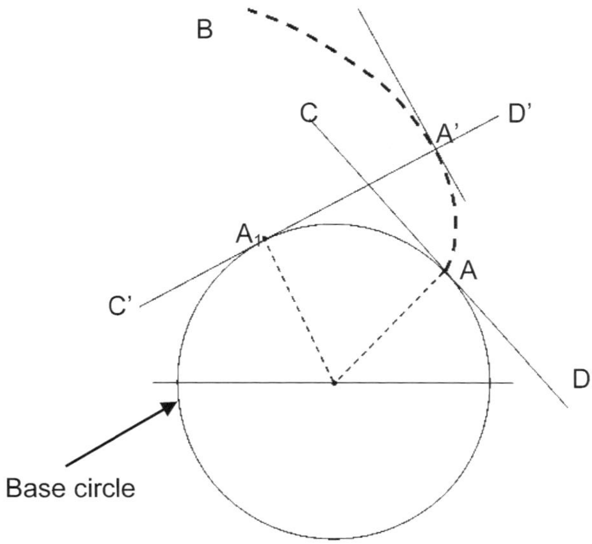

Fig. 6.9 Involute

Conjugate Profiles

Two mating profiles that satisfy the law of gear tooth action.

6.5 Involute as a Gear Tooth Profile

Involute

Curve $(AA'B)$ traced by a point (A) on a straight line (CD) as it rolls without slipping on a base circle, see Figure 6.9. Note the following:

- Arc distance $AA_1 = A_1A'$.
- Instantaneous center of rotation of the line at any instant $C'D'$ is at the contact point A_1 touching the base circle.
- The generating line $C'D'$ itself is the normal to the involute at any instant.
- Consequently, any normal to an involute is tangential to the base circle.

An involute profile can be easily generated by a trace point T on a tight cord by unwrapping it from the base cylinder as shown in Figure 6.10. Join the lines OT_0, OT and OA and denote the angles by $\angle AOT = \varphi$ and $\angle TOT_0 = \alpha$. Let the radius of the base cylinder be r_b, then

$$AT_0 = r_b\,(\varphi + \alpha) = AT = r_b \tan\varphi$$

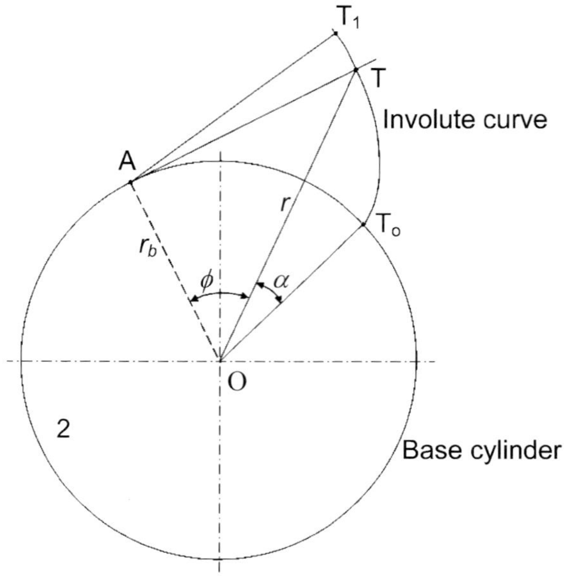

Fig. 6.10 Involute of a gear tooth profile

Therefore,

$$\alpha = \tan\varphi - \varphi = \text{inv}\,\varphi$$

and

$$OT = r = \frac{r_b}{\cos\varphi} \tag{6.7}$$

Involute as a Tooth Profile

Choose two base cylinders and wrap a cord tightly as shown in Figure 6.11a. Choose a tracing point T and rotate the cylinders slowly (keeping the cord always tight) to trace two involute curves *CD* and *EF*. Now, imagine two templates cut to the profiles *CD* and *EF* and fix them to the two base cylinders as shown in Figure 6.11b, to occupy the previous position of the traced involutes. You now have two conjugate profiles that can transmit motion at constant angular velocity ratio. Note that *AB* is the common normal to the involute profiles and it always intersects the line of centers at point P, which is the pitch point. The circles passing through the pitch point are pitch circles.

Path of Contact

Path traced by the contact point.

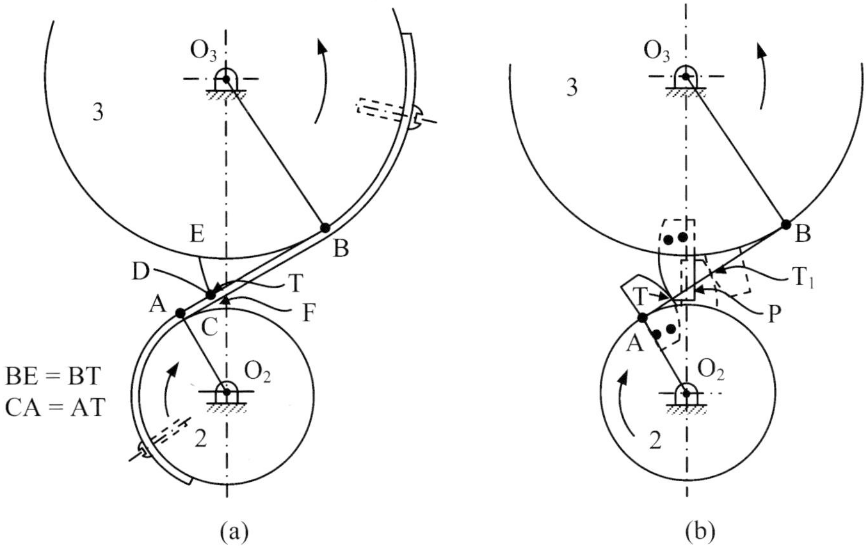

Fig. 6.11 Verification of the law of gear tooth action

Line of Action

Normal to both the mating profiles at the contact point.

Pressure Line

Path of contact being normal to the involute profiles is the pressure line.

Pressure Angle

Angle made by the pressure line with the horizontal.

6.6 Layout of an Involute Gear Set

As an example, draw the profile of the involute of the gear wheel and layout of the gear with the following specifications. The pinion 2 of 5 cm diameter and 4-diametral pitch mates with a gear wheel 3 with 50 teeth. The pressure angle is 20°.

- $d_3 = T_3/p_d = 50/4 = 12.5$ cm.
- Number of teeth on pinion $t_2 = d_2 \times p_d = 5 \times 4 = 20$.

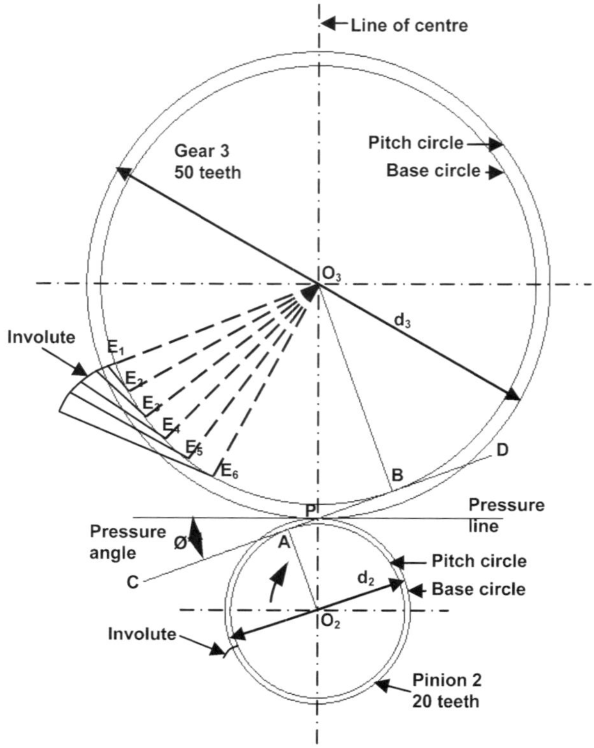

(a)

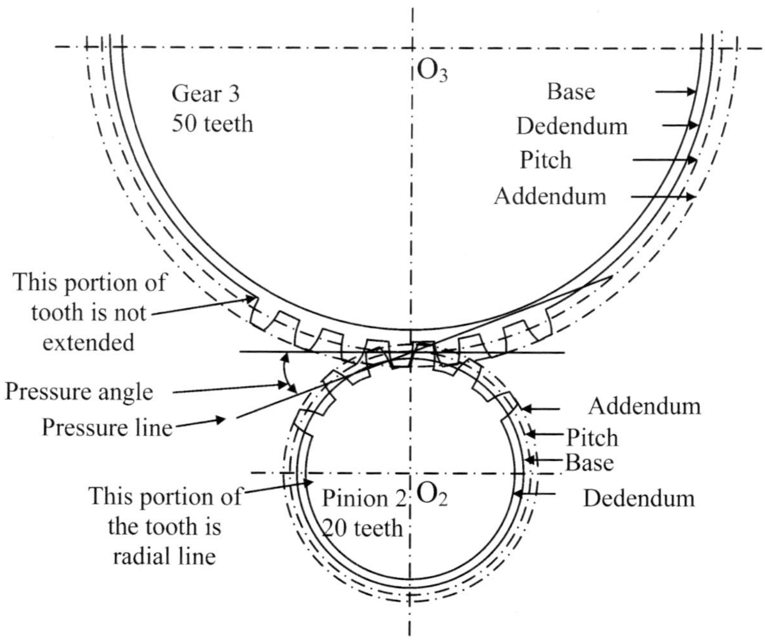

(b)

Fig. 6.12 Layouts of involute gear set

- Construct the pitch circles of diameters 5 and 12.5 cm respectively to touch at point P as shown in Figure 6.12a.
- Draw the pressure line inclined at $\varphi = 20°$ to the horizontal.
- Drop perpendiculars O_2A and O_3B on the pressure line to represent the base circle radii of the pinion and the wheel respectively.
- Draw the two base circles with O_2A and O_3B as radii.
- To generate the gear wheel involute, choose a convenient starting point O_3E_1.
- Construct a suitable number (say 6) of equal angle, say 10°, radial lines O_3E_1 to O_3E_6. The arc distance on the base circle of the wheel between the successive radial lines is given by $12.5 \times 10 \times \pi/180 = 2.18166$ cm. (Use any software and make these divisions smaller to get very accurate profile.)
- Draw perpendicular lines to each of these radial lines.
- Mark points on each of these perpendicular lines consecutively at distances $2.18166, 4.363323, 6.544983, \ldots$ beginning from the radial line position O_3E_2. Draw a smooth curve passing through these points to give the gear wheel involute tooth profile.
- Cut templates corresponding to the tooth profile thus drawn and the opposite side of the profile by turning over the previous template.
- $p_c = \pi/p_d = \pi/4 = 0.7853$ cm. Therefore the width of the tooth space is 0.39265 cm.
- Layout points which are 0.39265 cm apart on pitch circles of the gear wheel. (or simply draw radial lines at $360/(2 \times 50) = 3.6°$).
- The standard addendum is one module, i.e., $1/4 = 0.25$ cm and the dedendum is 1.157 times the module, i.e., $1.157/4 = 0.28925$ cm. Draw the addendum and dedendum circles.
- Using the templates cut earlier; trace the involute profile through the points marked on base circles.
- In the clearance portion of the gear, the involute may be rounded to form a fillet with the dedendum circle.
- Repeat the above process for the pinion.
- In this case, for the pinion, the dedendum circle is below the base circle and no involute portion exists in this region, see Figure 6.12b. In such cases, the involute profile is extended by a radial line and a fillet formed with the dedendum circle. This causes interference in gears which is discussed later.

Interference

Mating of a pinion with gear whose teeth profiles are not conjugate.

The construction of an internal gear or annulus driven by a pinion is shown in Figure 6.13. The internal gear tooth profile is concave, instead of convex for an external gear. Here, interference can occur if the base circle is larger than the addendum circle, i.e., a portion of the profile on the face is not an involute, unlike in an external gear where a portion of the tooth profile on the flank is not an involute. The

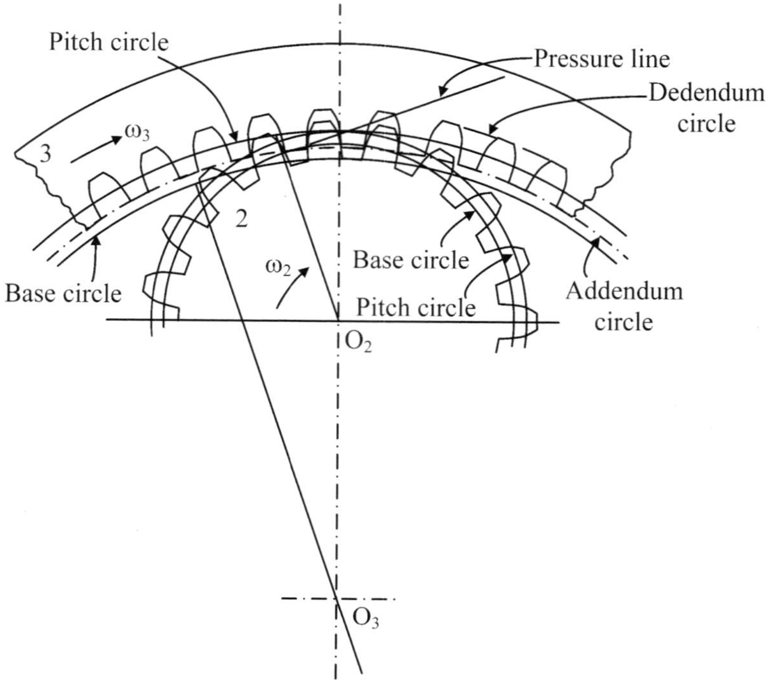

Fig. 6.13 Involute profile for internal gears

base circle of an internal gear should therefore be smaller than the addendum circle as shown in Figure 6.13.

Influence of Center Distance on Speed Ratio

The speed ratio of the gear pair is given by $\omega_3/\omega_2 = O_2C/O_3C$, see equation (6.5). Since triangles O_2AP and O_3BP are similar, the speed ratio is

$$\frac{\omega_3}{\omega_2} = \frac{O_2A}{O_3B} \tag{6.8}$$

Therefore, the velocity ratio is also equal to the inverse ratio of the base circle radii. This is a great advantage for the involute tooth profile that the center distance of the gears need not be very exact, as far as the speed ratio is concerned, since the base circle radii do not change with mounting inaccuracies.

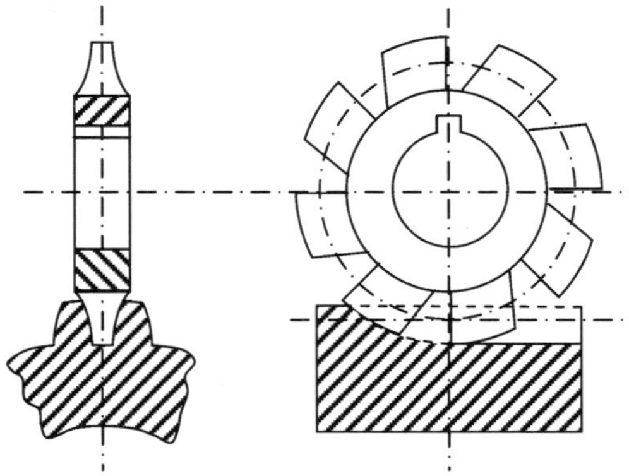

Fig. 6.14 Forming process

6.7 Producing Gear Teeth

Forming Process

A milling cutter, so shaped that the required tooth outline is obtained by setting the cutter to the desired depth, is used for machining the material in the space of gear blank on a milling machine, see Figure 6.14.

Generating Process

Correct profiles are produced by making one of the gears of a set a cutting tool and the other a blank. The pitch circle of the cutter is forced to roll with the pitch circle of a gear blank as though they are mating with each other.

Undercutting

If conditions similar to the pinion of Figure 6.13 exist, where the tooth profile extends below the base circle (interference of the tooth profile), the generating cutter removes the interfering portion of the blank. This process is called undercutting, which automatically avoids interference in the gears; however, it weakens the tooth at the root.

The generating process is usually accomplished by three types of cutters: (1) hob, (2) pinion shaped cutter, and (3) rack shaped cutter.

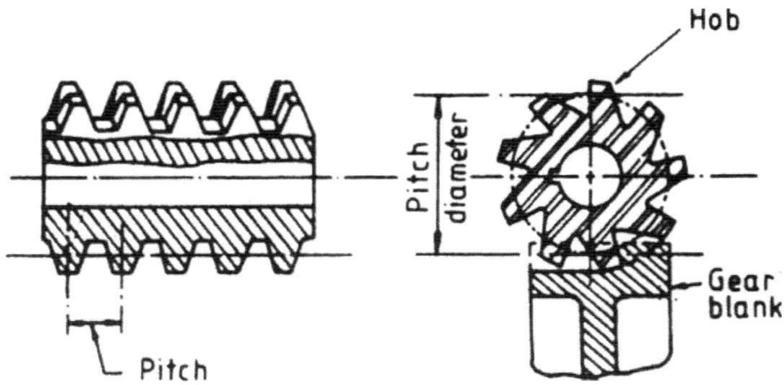

Fig. 6.15 Hobbing process

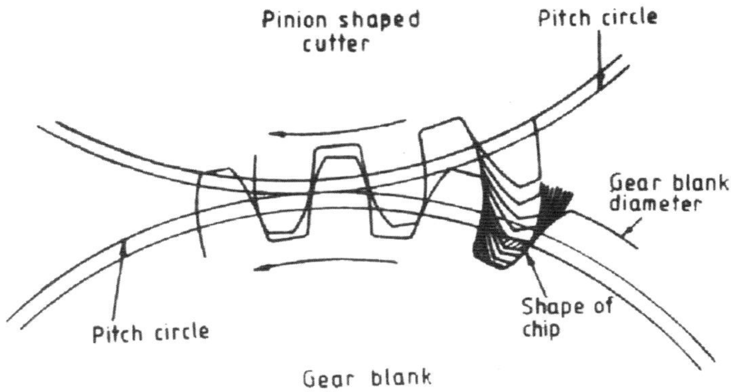

Fig. 6.16 The generating process of a gear tooth shown by a pinion shaped cutter

Hobbing

In the hobbing process, a hob resembling a worm-shaped cutter is advanced through a gear blank when both are rotating as shown in Figure 6.15.

Generating by Pinion Shaped Cutter

Both the cutter as well as the blank are rotated, as if they are in mesh, with a small feed for every revolution of the blank, see Figure 6.16.

Generating by Rack Shaped Cutter

Most accurate method of generating process, since an involute rack has straight teeth, see Figure 6.17.

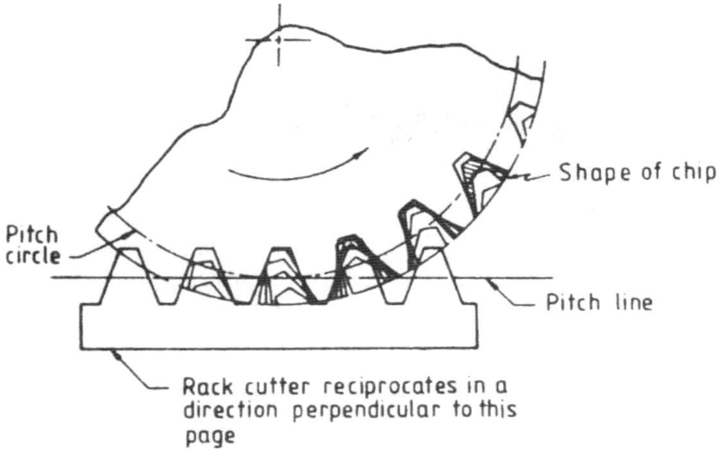

Fig. 6.17 The process of cutting a gear by the use of a rack shaped cutter

6.8 Meshing Gears and Line of Contact

Consider a gear pair with 40 and 20 teeth as shown in Figure 6.18a The addendum and dedendum taken are 6.366 and 7.957 mm with pressure angle 25°. Center distance is 190.985 mm.

A close-up view of meshing teeth is shown in Figure 6.18b. In this position the gear tooth tip is in contact with the pinion tooth flank where the engagement begins at point A on the pressure line and the disengagement at point B. AB is the path of contact in the meshing process of the gear tooth pair. A further close-up of engagement and disengagement is shown in Figure 6.18c.

6.9 Interference of Involute Gears

Figure 6.19 shows two mating teeth of a driving and driven gear. Here, the dedendum circles of both the wheels are below their base circles. When mating takes place, the first contact is at point A and the non-involute flank portion of the driving gear transmits motion with the involute face portion of the driven gear and interference occurs. When the gear tooth pair is disengaged from the transmission of motion, interference on the flank of the driven gear takes place as shown in Figure 6.19. O_2C and O_3D are perpendiculars dropped onto the pressure line. AB is the total path of contact out of which only CD has correct tooth action.

To avoid interference two methods can be used.

1. Increase the pressure angle, so that A and B lie within the maximum allowed path of contact CD.

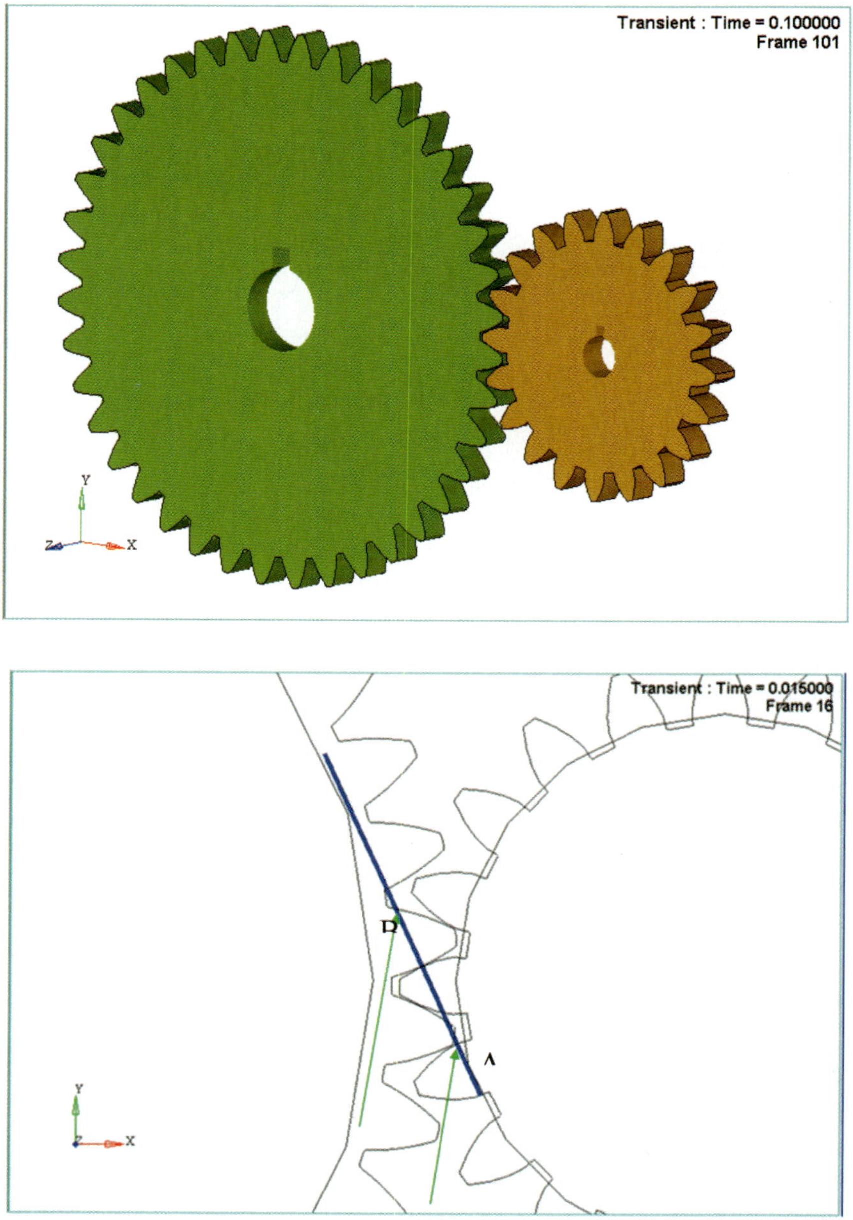

(a)

(b)

Fig. 6.18 (a) Mating gears. (b) Close-up view of meshing teeth and line of contact

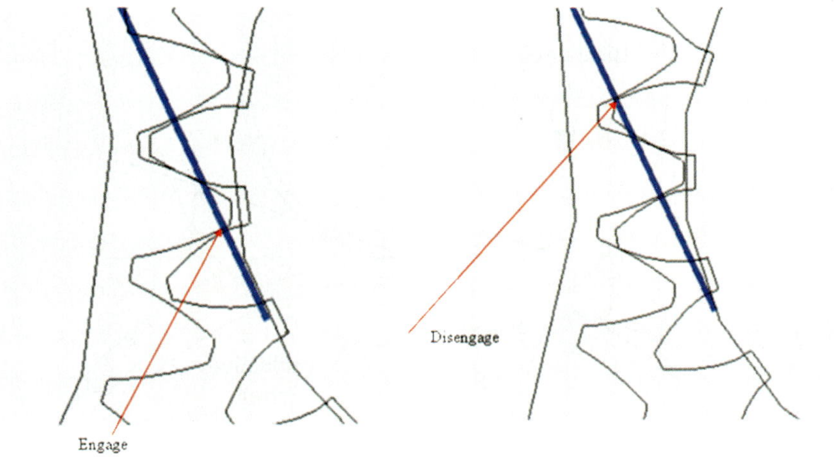

(c)

Fig. 6.18 (Continued) (c) Close-up view engagement and disengagement in meshing

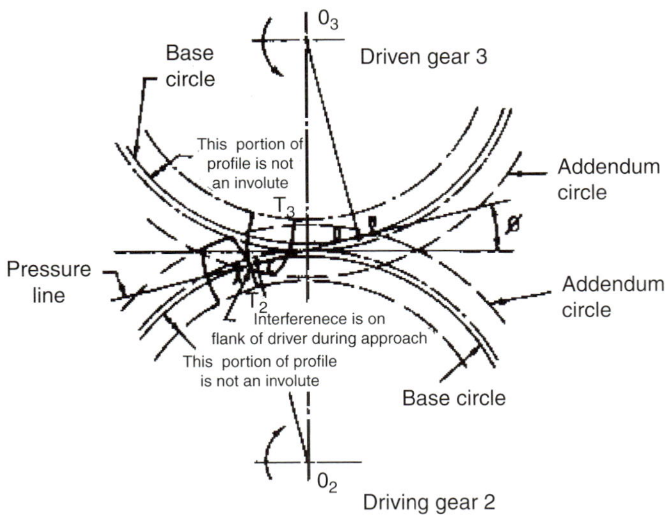

Fig. 6.19 Two mating teeth of a pair of gears

2. Decrease the addendum and dedendum of the gear wheel, i.e., increase the diametral pitch which means increase the number of teeth.

Since the pinion has a lesser number of teeth, there should be a minimum number of teeth on the pinion for a given pressure angle to avoid any interference.

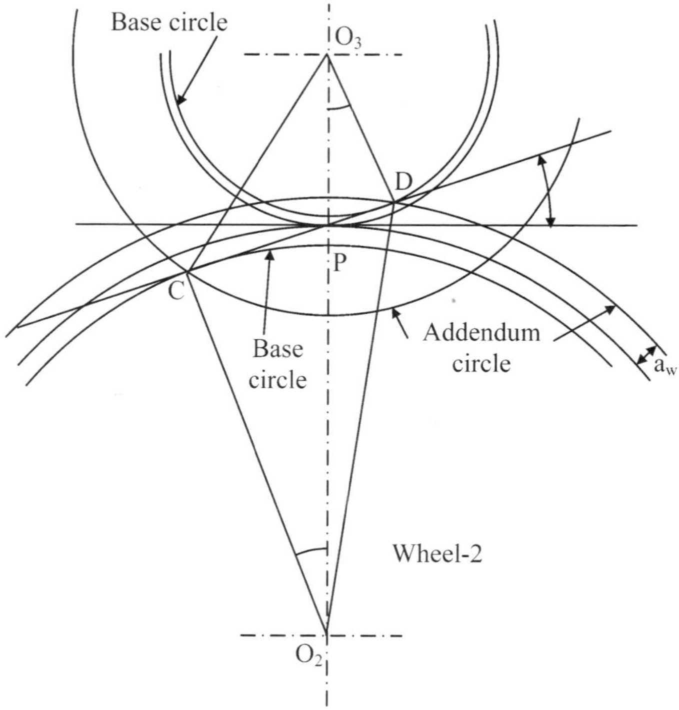

Fig. 6.20 Addendum circles of a pinion and a wheel

6.10 Minimum Number of Teeth to Avoid Interference

First draw the pitch and base circles of a mating pair of wheels, the pinion 3 and gear 2. Also, draw the pressure line and drop the perpendiculars O_2C and O_3D from the gear and pinion centers as shown in Figure 6.20. The involute profile can exist for a path of action beginning from C and ending at D, therefore draw the addendum circle for the pinion through C and for gear through D. Since the addenda for both the pinion and the wheel are same, the governing addendum is that of the wheel, a_w.

From the triangle O_2PD

$$\begin{aligned} O_2D^2 &= O_2P^2 + DP^2 - 2 \times O_2P \times PD \times \cos O_2PD \\ &= O_2P^2 + DP^2 + 2 \times O_2P \times PD \times \sin \varphi \end{aligned} \tag{6.9}$$

From the triangle O_3PD

$$PD = O_3P \times \sin \varphi \tag{6.10}$$

Therefore equation (6.9) can be written as

$$O_2D^2 = O_2P^2 + O_3P^2\sin^2\varphi + 2 \times O_2P \times O_3P \times \sin^2\varphi$$
$$= O_2P^2\left[1 + \frac{O_3P}{O_2P}\left(\frac{O_3P}{O_2P} + 2\right)\sin^2\varphi\right] \quad (6.11)$$

Or

$$O_2D = O_2P\left[1 + \frac{O_3P}{O_2P}\left(\frac{O_3P}{O_2P} + 2\right)\sin^2\varphi\right] \quad (6.12)$$

The addendum of the wheel is $a_w = O_2D - O_2P$, which from the above equation (6.12) can be expressed as

$$a_w = O_2P\left[\sqrt{1 + \frac{O_3P}{O_2P}\left(\frac{O_3P}{O_2P} + 2\right)\sin^2\varphi} - 1\right] \quad (6.13)$$

Let t be the number of pinion teeth, T be the number of teeth on the wheel, then $O_2P = T/2p_d$ and $O_3P/O_2P = t/T$. Therefore, equation (6.13) becomes

$$a_w = \frac{T}{2p_d}\left[\sqrt{1 + \frac{t}{T}\left(\frac{t}{T} + 2\right)\sin^2\varphi} - 1\right] \quad (6.14)$$

Or

$$T = \frac{2a_w p_d}{\left[\sqrt{1 + \frac{t}{T}\left(\frac{t}{T} + 2\right)\sin^2\varphi} - 1\right]} \quad (6.15)$$

Therefore, the minimum number of teeth on the pinion to avoid interference is given by

$$t_{\min} = \frac{2a_w \frac{t}{T} p_d}{\left[\sqrt{1 + \frac{t}{T}\left(\frac{t}{T} + 2\right)\sin^2\varphi} - 1\right]} \quad (6.16)$$

For a rack and pinion, the corresponding figure is given in Figure 6.21. Here the addendum of the rack, a_r, is the governing value and should not exceed DV.

$$DV = PD\sin\varphi \quad (6.17)$$

Also,

$$PD = O_3P\sin\varphi \quad (6.18)$$

Therefore,

$$DV = a_r = O_3P\sin^2\varphi \quad (6.19)$$

Further, $O_3P = t/2p_d$ and hence

$$a_r = \frac{t}{2p_d}\sin^2\varphi \quad (6.20)$$

Or

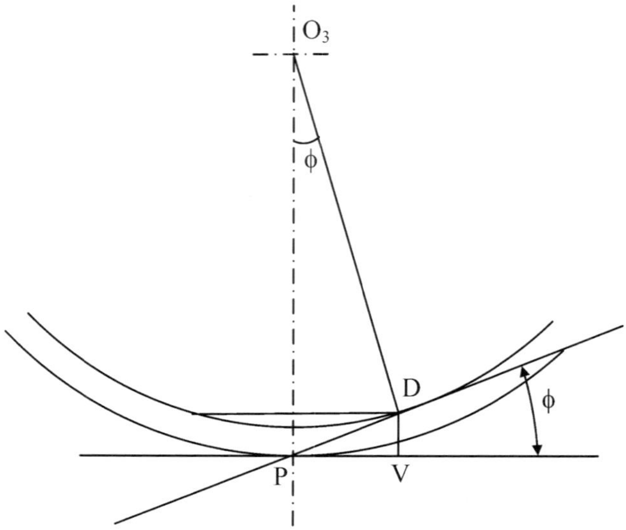

Fig. 6.21 A pinion mating with a rack

$$t_{\min} = \frac{2a_r p_d}{\sin^2 \varphi} \tag{6.21}$$

6.11 Contact Ratio

Arc of Approach

Arc distance on the pitch circle from the point of engagement to pitch point.

Arc of Recess

Arc distance on the pitch circle from the pitch point to point of disengagement.

Arc of Action

Sum of arc of approach and arc of recess.

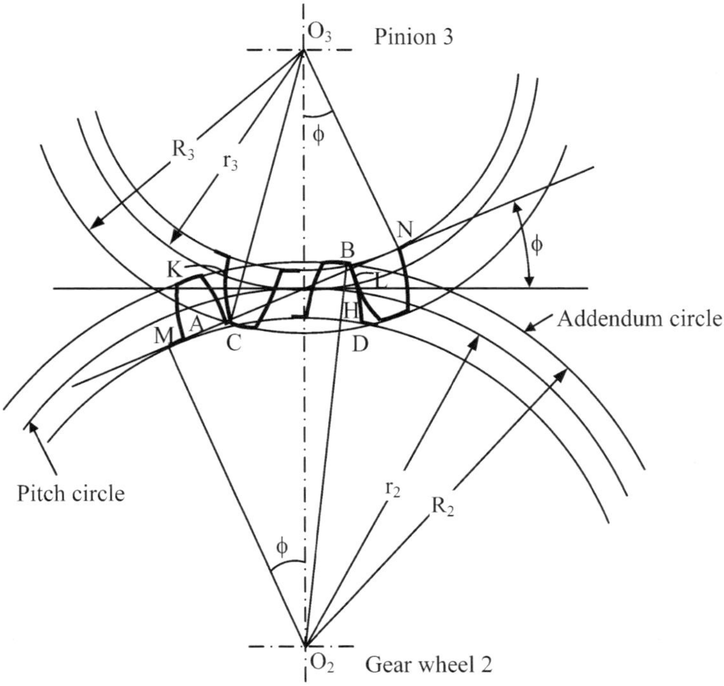

Fig. 6.22 Engaging and disengaging positions of a tooth pair

Path of Approach

Length measured along the pressure line from the point of engagement to the pitch point.

Path of Recess

Length measured along the pressure line from the pitch point to the point of disengagement.

Path of Contact

Sum of the path of contact and recess.

Figure 6.22 shows a pair of teeth in contact without any interference. The arc distance *GP* on the pitch circle of gear wheel 2 or *KP* on the pitch circle of pinion 3 is the arc of approach. Similarly, the arc distance *PH* or *PL* on pitch circles of gear wheel 2 and pinion 3 respectively is the arc of recess. The total arc distance *GH* or *KL* is the arc of action.

Contact Ratio

The ratio of arc of action to circular pitch. Obviously, this should be more than 1, as otherwise there will be no new pair coming into action after one pair has completed its job.

The length of path of contact is

$$AB = AP + PB \tag{6.22}$$

Also,

$$AP = AN - PN = \sqrt{R_3^2 - r_3^2\cos^2\varphi} - r_3\sin\varphi \tag{6.23}$$

and

$$PB = BM - PM = \sqrt{R_2^2 - r_2^2\cos^2\varphi} - r_2\sin\varphi \tag{6.24}$$

Substituting the above two equations (6.23 and (6.24) in (6.22), the total path of contact is

$$AB = \sqrt{R_3^2 - r_3^2\cos^2\varphi} + \sqrt{R_2^2 - r_2^2\cos^2\varphi} - (r_2 + r_3)\sin\varphi \tag{6.25}$$

From the properties of the involute, we have the arc *CD* on the base circle equal to the path of contact *AB* given in equation (6.25). Therefore,

$$\text{Arc } CD = \sqrt{R_3^2 - r_3^2\cos^2\varphi} + \sqrt{R_2^2 - r_2^2\cos^2\varphi} - (r_2 + r_3)\sin\varphi \tag{6.26}$$

From Figure 6.22 we note that $O_2P = O_2M/\cos\varphi$ (see also equation 6.7) is the property of involute. Therefore the arc distance *GH* on the pitch circle is equal to arc distance *CD* on the base circle divided by $\cos\varphi$. Hence,

$$\text{Arc of Approach} = \frac{\sqrt{R_3^2 - r_3^2\cos^2\varphi} - r_3\sin\varphi}{\cos\varphi} \tag{6.27}$$

$$\text{Arc of Recess} = \frac{\sqrt{R_2^2 - r_2^2\cos^2\varphi} - r_2\sin\varphi}{\cos\varphi} \tag{6.28}$$

$$\text{Arc of Contact} = \frac{\sqrt{R_3^2 - r_3^2\cos^2\varphi} + \sqrt{R_2^2 - r_2^2\cos^2\varphi} - (r_2 + r_3)\sin\varphi}{\cos\varphi} \tag{6.29}$$

and the contact ratio is

$$\text{Contact Ratio} = \frac{\sqrt{R_3^2 - r_3^2\cos^2\varphi} + \sqrt{R_2^2 - r_2^2\cos^2\varphi} - (r_2 + r_3)\sin\varphi}{p_c\cos\varphi} \tag{6.30}$$

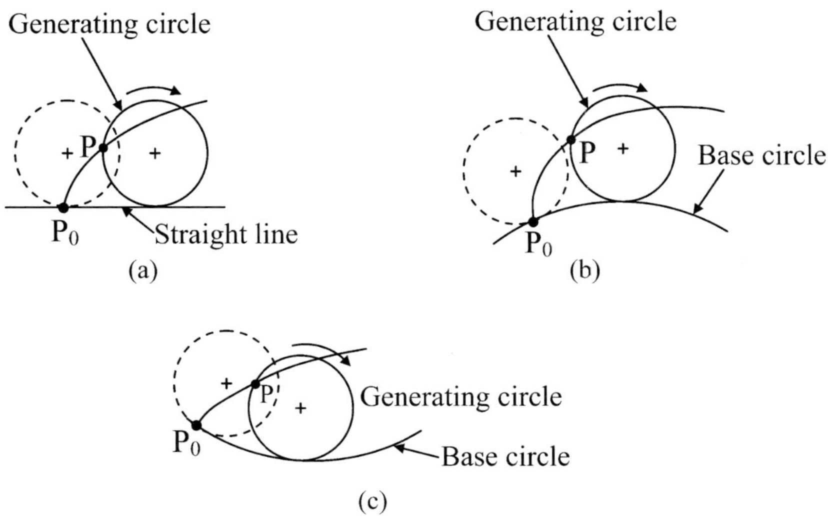

Fig. 6.23 Cycloidal tooth profiles

6.12 Cycloidal Tooth Profiles

Cycloid

Curve described by a point on the circumference of a circle (generating circle) as it rolls without slipping on a straight plane, see Figure 6.23a.

Epicycloid

The path traced by a point on the generating circle as it rolls on another circle (base circle), see Figure 6.23b.

Hypocycloid

The path traced by a point on the generating circle as it rolls on the inside of base circle, see Figure 6.23c.

Generation of Cycloidal Gear Teeth

Figures 6.24a and b show the generation of cycloidal gear teeth on the gear wheel 3 and pinion 2 respectively.

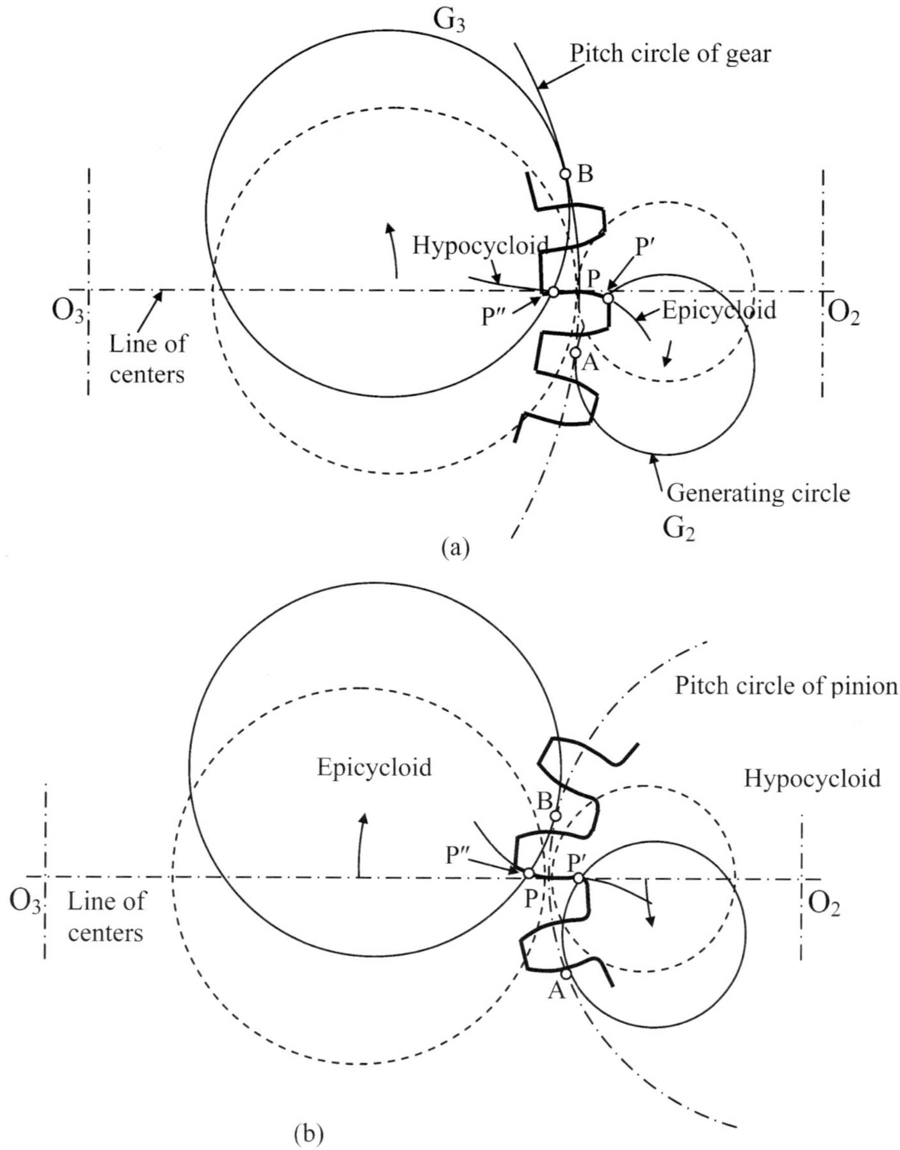

Fig. 6.24 (a) Generation of cycloidal tooth on gear and (b) pinion

- Draw the pitch circle of the gear. This will serve as the base circle for generation of a cycloid tooth form.
- Choose a generating circle $G2$ (usually smaller than the pitch circle of the gear). Draw this circle with its center lying on the line of centers of the gear pair and in the initial position touching the pitch circle of gear 3 at the pitch point P.

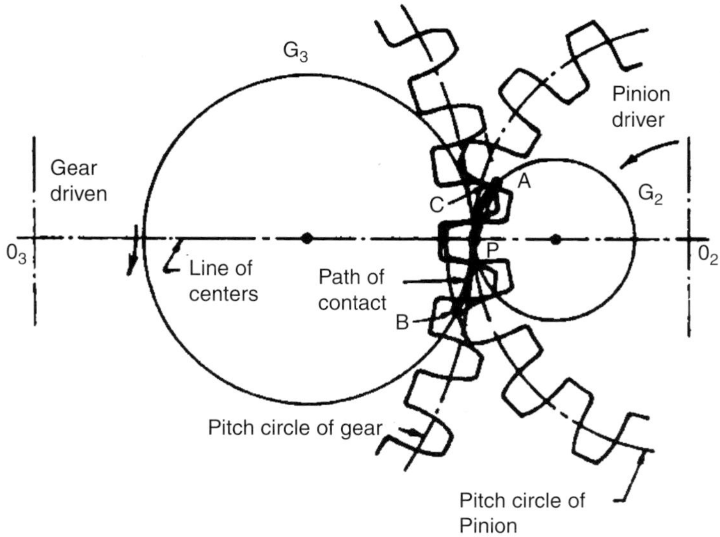

Fig. 6.25 Generated pinion and gear

- Roll the generating circle $G2$ clockwise on the pitch circle ($AP = AP'$) to generate the face of the tooth by an epicycloid obtained from P on the generating circle.
- Choose a generating circle $G3$ (usually larger than the generating circle $G2$, but smaller than the base circle). Draw this circle with its center lying on the line of centers of the gear pair and in the initial position touching the pitch circle of gear 3 at the pitch point P.
- Roll the generating circle $G3$ counter-clockwise on the pitch circle ($BP = BP''$) to generate the flank of the tooth by a hypocycloid obtained from P on the generating circle.
- Follow the same procedure to draw the tooth profile for the pinion. Remember the same generating circles should be used.
- You can make templates of these profiles and the ones turned over and use them to complete the tooth profiles all around the pitch circle at appropriate angular distances depending on the circular pitch of the gear pair.
- Draw the addendum and dedendum circles and complete the formation of tooth profiles.

Figure 6.25 shows the pinion and gear with cycloidal teeth in mesh. When the two pitch circles rotate without slipping, the generating circles will also roll over each other simultaneously without slipping. If the pinion is the driver, the engagement starts at point A, the intersection of addendum circle of the gear wheel and the generating circle $G2$, and the disengagement takes place at point B which is the intersection of addendum circle of the pinion with the generating circle $G3$.

The path of contact is *APB*. The common normal to the mating profiles at any instant always passes through the pitch point. However, the common normal makes different angles during the path of contact with a maximum value at the engagement and disengagement positions and zero at the pitch point. Therefore, the pressure angle in cycloidal gears is not constant, as in the case of involute gears. This variation of pressure angle leads to noise in cycloidal gears which is not present in involute gears. However, the main advantage of cycloidal gears is that the question of interference does not arise as in the case of involute gears.

6.13 Cycloidal and Involute Tooth Forms

1. The involute profile is single and simple in nature, unlike the epicycloid and hypocycloid combination forming a tooth in cycloidal gears.
2. The involute is a straight line for the case of a rack, which is the fundamental form of cutting tools. Therefore, the generation of involute gears is very accurate and less expensive.
3. The pressure angle of involute gears is constant, making them less noisy when compared with cycloidal gears which have a varying pressure angle.
4. In the case of involute gears, their center distance need not be very accurately maintained as shown in equation (6.8), thus allowing a slight change in the center distance while mounting the gears and shafts.
5. The main advantage of cycloidal gears is that there is no possibility of interference to occur, provided the center distance is accurately maintained.

In view of a large number of advantages of the involute profile, it has superseded the cycloidal profile to the extent that no standard systems for interchangeable gears exist for cycloidal profiles.

6.14 Solved Problems

Solved Problem 6.1

A 2-DP, 24 tooth pinion drives a 36 tooth gear using 14.5° full depth involute system layout of the gears. Determine the arcs of approach, recess and contact ratio of the gear pair.

1. Pitch radius of pinion $r_3 = 1/2 \times 24/2 = 6$ cm.
2. Pitch radius of the gear $r_2 = 36/24 \times 6 = 9$ cm.
3. Pinion base circle radius $6 \times \cos 14.5 = 5.808886$ m.
4. Gear base circle radius $9 \times \cos 14.5 = 8.713329$.
5. Use standard addendum $a = 1/p_d = 0.5$. Therefore $R_3 = 6.5$ and $R_2 = 9.5$ cm.

6. Draw the pitch circles, base circles and addendum circles of the gear and pinion as shown in Figure 6.26.
7. Draw the pressure line.
8. Drop perpendiculars from gear centers on to the pressure line and mark the points M and N as shown in figure.
9. Also mark the points A and B, the intersection points of pressure line with the addendum circles.
10. Measure the path of approach $AP = 1.4161$ cm.
11. Determine the arc of approach $1.4161/\cos 14.5 = 1.4627$ cm.
12. From (6.23)

$$
\begin{aligned}
AP &= \sqrt{R_3^2 - r_3^2\cos^2\varphi} - r_3 \sin\varphi \\
&= \sqrt{6.5^2 - 6^2\cos^2 14.5} - 6 \sin 14.5 \\
&= 1.41437
\end{aligned}
$$

13. Therefore the calculated values of arc of approach $1.41437/\cos 14.5 = 1.4609$ cm.
14. Repeat steps 10 to 13 above for the arc of recess to get the value from construction in Figure 6.26 as 1.5804 cm and from equation (6.24) as 1.5820 cm.
15. Total arc of contact is $1.4627 + 1.5804 = 3.0431$ cm.
16. Circular pitch $p_c = \pi/2 = 1.5708$ cm.
17. Therefore, the contact ratio is $3.0431/1.5708 = 1.9373$.

Solved Problem 6.2

A pinion of 20 teeth is to drive an internal wheel of 64 teeth. Using 20° full depth involute system and 2.5 DP, make a drawing of the pinion tooth profile.

1. Pinion pitch circle radius $r_3 = 1/2 \times 20/2.5 = 4$ cm.
2. Pinion base circle radius $4 \cos 20 = 3.75877$ cm.
3. Use standard addendum $a = 1/p_d = 0.4$. Therefore $R_3 = 4.4$ cm.
4. Standard dedendum$d = 1.157/p_d = 0.4628$. Therefore dedendum circle radius is 3.5372 cm.
5. Draw the various circles as shown in Figure 6.27.
6. Use 5° sector positions (seven used in the figure) on the base circle and draw the radial lines, starting from position 0 at a convenient location to position 7.
7. At these radial line positions, draw tangents as shown.
8. Mark on each tangent, distances $4 \times 5 \times \pi/180 = 0.349066$ cm and tangent line till the seventh tangent line and join these points by a smooth curve to give the tooth profile. Note that the profile is a radial line below the base circle up to the dedendum circle.

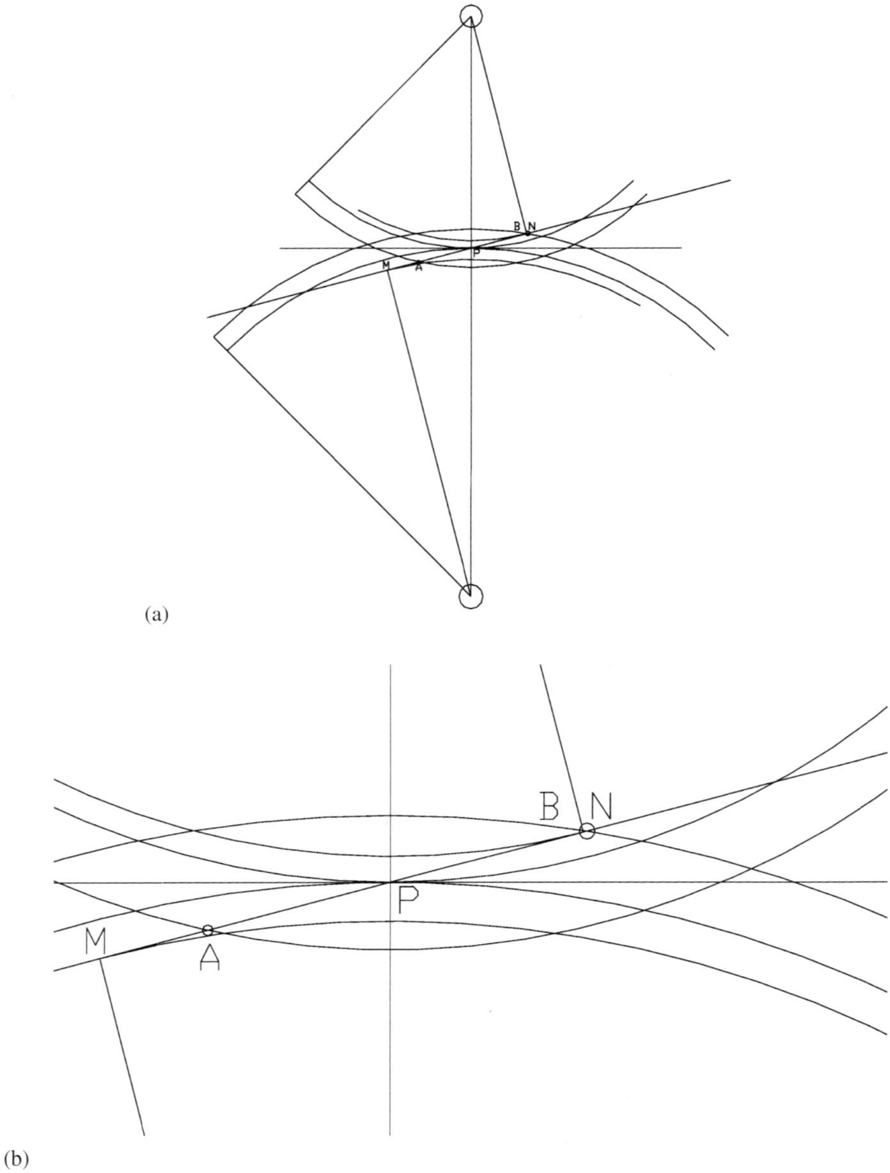

Fig. 6.26

Solved Problem 6.3

Layout a 2-DP 24 tooth gear meshing with a rack, using 20° full depth involute system. Determine the contact ratio of this rack and pinion set.

1. Pitch radius of pinion $r_3 = 1/2 \times 24/2 = 6$ cm.

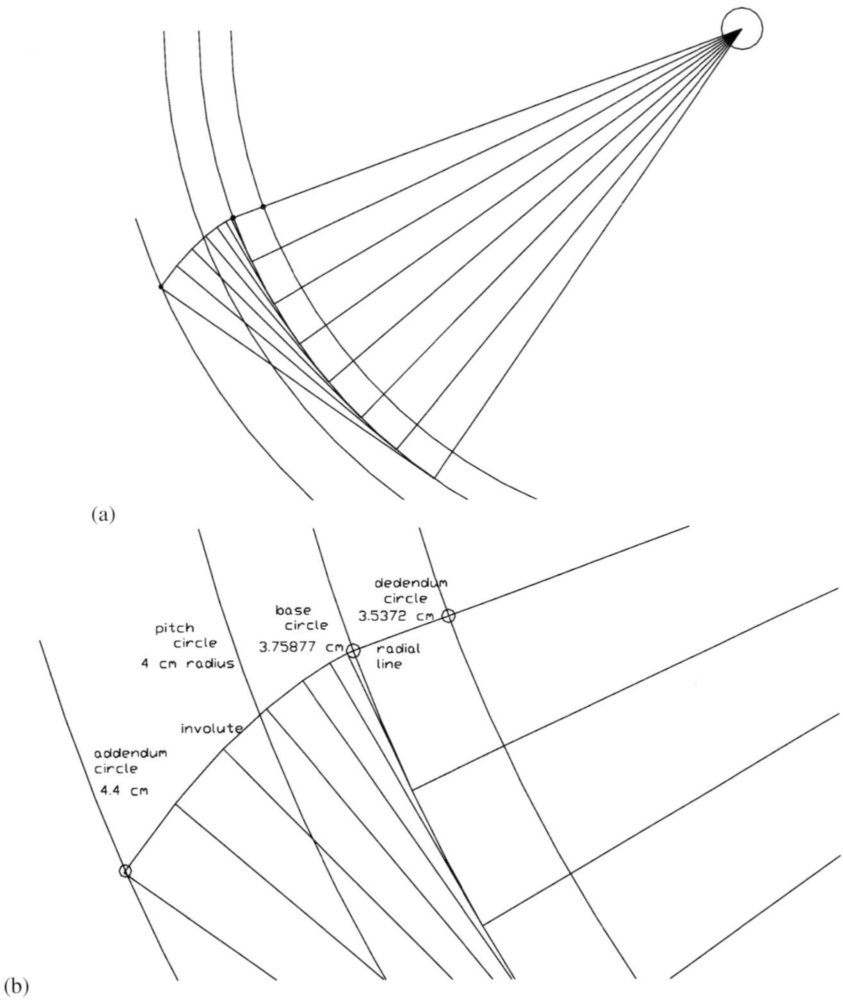

Fig. 6.27

2. Pinion base circle radius $r_2 = 6\cos 20 = 5.638156$ cm.
3. Addendum $a = 1/p_d = 0.5$. Therefore $R_3 = 6.5$ Cm.
4. Dedendum $d = 1.157/p_d = 0.5785$. Therefore dedendum circle radius is 5.4215 cm.
5. Draw all the pinion circles as shown in Figure 6.28.
6. Draw a horizontal line touching the point P to represent the pitch line of the rack.
7. Draw parallel lines to the rack pitch line at distances 0.5 cm above and 0.5785 cm below to represent the rack addendum and dedendum lines.
8. Draw the pressure line MN.
9. AB is the path of contact, measure this value which is 2.6442 cm.

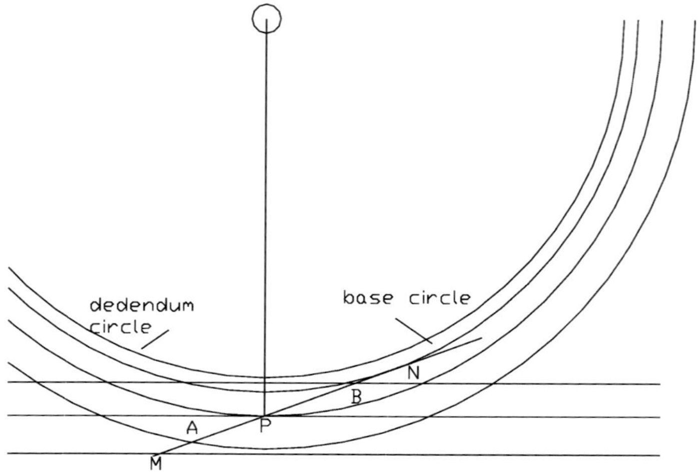

Fig. 6.28

10. Arc of contact is $2.6442/\cos 20 = 2.8139$ cm.
11. Circular pitch $p_c = \pi/p_d = 1.5707$ cm.
12. Contact ratio is $2.8139/1.5707 = 1.79$.

Solved Problem 6.4

Draw the profile of a cycloidal rack tooth mating with a pinion having 16 teeth. The circular pitch is 6 cm and the rolling circle for the face as well as flank has a diameter equal to the pitch circle diameter of a 12 tooth pinion.

1. Addendum $a = 1/p_d = p_c/\pi = 1.909859$ cm.
2. Dedendum $d = 1.157a = 2.209707$ cm.
3. Generating circle diameter $p_c \times T/\pi = 6 \times 12/\pi = 22.9183$ cm.
4. Draw the addendum, pitch and dedendum lines of the rack and the generating circles as shown in Figure 6.29.
5. Roll the generating circles on the pitch line of the rack in opposite directions in equal intervals, following the procedure outlined in Section 6.12 to obtain the cycloidal tooth of the rack.

Solved Problem 6.5

What is the smallest number of teeth that can be used on each of two equal gears to avoid interference, if the pressure angle is 14.5°?

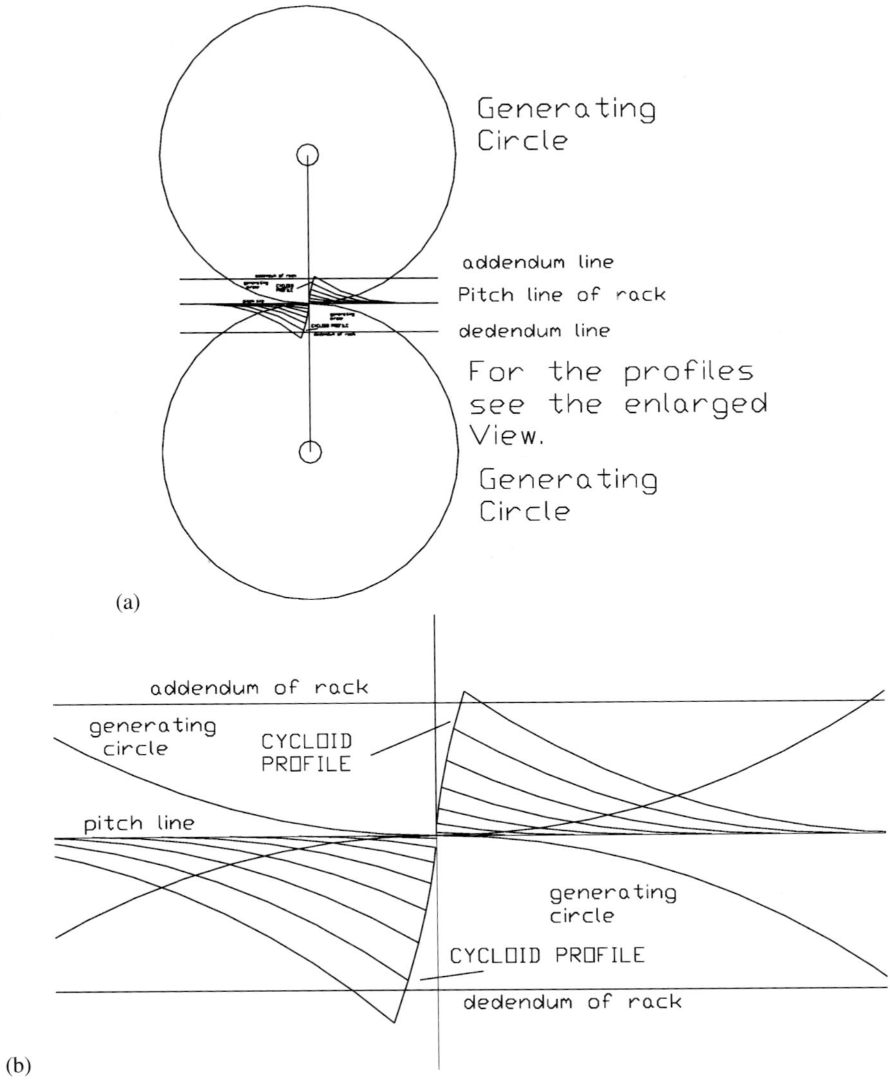

Fig. 6.29

1. From equation (6.16), we have

$$t_{\min} = \frac{2a_w \frac{t}{T} p_d}{\left[\sqrt{1 + \frac{t}{T}\left(\frac{t}{T} + 2\right)\sin^2\varphi} - 1\right]}$$

2. Use standard value for addendum $a_w = 1/p_d$.
3. $t/T = 1$.

4. Therefore,

$$t_{\text{min}} = \frac{2}{\left[\sqrt{1 + 3\sin^2 14.5} - 1\right]}$$
$$= 22.226$$
$$= 23$$

Solved Problem 6.6

Determine the addendum of the teeth of a gear pair consisting of two spur wheels each having 30 teeth to have a minimum contact ratio equal to 2. The circular pitch is 2.5 cm and the pressure angle is 20°.

1. Contact ratio is

$$\frac{\sqrt{R_3^2 - r_3^2\cos^2\varphi} + \sqrt{R_2^2 - r_2^2\cos^2\varphi} - (r_2 + r_3)\sin\varphi}{p_c \cos\varphi}$$

2. Since both the gears are of same size, the contact ratio is

$$\frac{2}{2.5\cos 20}\left[\sqrt{R_2^2 - r_2^2\cos^2 20} - r_2 \sin 20\right] = 2$$

3. The pitch radius is given by $r_2 = 30 \times 2.5/2\pi = 11.9366$ cm.
4. Therefore,

$$\sqrt{R_2^2 - 11.9366^2 \times 0.9397^2} = 0.342 \times 11.9366 + 2.34935$$

5. Squaring both sides $R_2^2 - 125.817 = 41.3651$.
6. Therefore $R_2 = 12.9299$ cm and the addendum is $12.9299 - 11.9366 = 0.9933$ cm.

Solved Problem 6.7

Two gear wheels 10 and 15 cm pitch diameters have involute teeth of 1.6 DP and pressure angle 20°. The addenda are 3 mm. Determine 1. Length of path of contact, 2. Contact ratio and 3. The angle turned through by the pinion while any one pair of teeth is in contact.

1. From equation (6.25)

$$AB = \sqrt{53^2 - 50^2 \cos^2 20} + \sqrt{78^2 - 75^2 \cos^2 20} - 125 \sin 20$$
$$= 15.195 \text{ mm}$$

2. Substituting known values, $r_3 = 50$ mm and $r_2 = 75$ mm and using given addenda 3 mm

$$AB = \sqrt{53^2 - 50^2 \cos^2 20} + \sqrt{78^2 - 75^2 \cos^2 20} - 125 \sin 20$$
$$= 15.195 \text{ mm}$$

3. Contact ratio is arc of contact divided by circular pitch. Arc of contact is path of contact divided by $\cos\varphi$.
4. Arc of contact is $1.5195/\cos 20$.
5. Circular pitch is π/p_d.
6. Therefore the contact ratio is $(1.5195/0.9396) \times 1.6/\pi = 0.8235$.
7. Angle turned by pinion while any one pair of teeth in contact is given by the arc of contact divided by the radius, i.e.,

$$\text{Angle turned by pinion} = \frac{\frac{1.5195}{0.93969}}{5} = 0.3234 \text{ rad}$$
$$= 18.53°$$

Solved Problem 6.8

Determine the minimum number of teeth that can be employed on the pinion to mesh with a rack, if the pressure angle is 14.5° and the addendum is one module. If the number of teeth is to be reduced from the value you find and that interference should be avoided, what changes would you recommend in the gear pair?

1. From equation (6.21)

$$t_{\min} = \frac{2a_r p_d}{\sin^2\phi} = \frac{2}{\sin^2 14.5} = 32$$

2. If the number of teeth is to be less than 32 as obtained above, the pressure angle can be increased to avoid interference. e.g., if the number of teeth allowed is 18, $\sin^2\phi = 2/18$, a pressure angle of 20° can be employed. If the number of teeth is to be further decreased to say 12, the pressure angle has to be increased to 25°.

Solved Problem 6.9

Determine the maximum addenda in terms of circular pitch of a pair of spur gears having 12 and 20 teeth, if undercutting is to be avoided. The pressure angle is 20°.

1. From equation (6.14)

$$a_w = \frac{T}{2p_d}\left[\sqrt{1+\frac{t}{T}\left(\frac{t}{T}+2\right)\sin^2\varphi}-1\right]$$

2. Substituting the known values,

$$a_w = \frac{20}{2\pi}\left[\sqrt{1+0.6\,(0.6+2)\sin^2 20}-1\right]p_c$$
$$= 0.2783p_c$$

3. Therefore the maximum addendum that can be used is 0.2783 times the circular pitch.

Solved Problem 6.10

Determine the minimum number of teeth on each wheel of an involute gear pair of speed ratio 3.5. The pressure angle is 14.5° and it is necessary that the arc of approach should be more than the circular pitch.

1. Equation (6.27) gives

$$\text{Arc of Approach} = \frac{\sqrt{R_3^2 - r_3^2\cos^2\varphi} - r_3\sin\varphi}{\cos\varphi}$$

2. Taking the arc of approach to be at least the circular pitch, we have the condition that

$$p_c\cos 14.5 = \sqrt{\left(r_3+\frac{p_c}{\pi}\right)^2 - r_3^2\cos^2 14.5} - r_3\sin 14.5$$

3. The relation between r_3, p_c and number of teeth t, is

$$r_3 = \frac{p_c t}{2\pi} = 0.15915p_c t$$

4. Therefore,

$$p_c\cos 14.5 =$$

$$\sqrt{\left(0.15915p_c t+\frac{p_c}{\pi}\right)^2 - (0.15915p_c t)^2\cos^2 14.5} - 0.15915p_c t\sin 14.5$$

5. Cancelling p_c

$$0.9681 = \sqrt{\left(0.15915t+\frac{1}{\pi}\right)^2 - (0.15915t)^2\times 0.9373} - 0.15915\times 0.2504t$$

6. The above gives $t = 34.6$.
7. To get the exact speed ratio, the nearest teeth numbers are $t = 36$ and $T = 126$.

6.15 Additional Problems

1. Determine the pitch diameter, circular pitch, addendum, dedendum, tooth thickness and clearance of a 22 tooth gear with standard full depth involute of diametral pitch 2.
2. A spur gear pair has pitch diameters of 12 and 30 cm. What is the largest tooth size, in terms of diametral pitch, that can be used without having any interference? Take the pressure angle to be 20°.
3. Two mating gear wheels have 20 and 40 involute teeth of diametral pitch 1 and 20° pressure angle. The addendum of each wheel is to be made of such a length that the line of contact on each side of the pitch point has half the maximum possible length. Determine the addenda and the length of arc of contact.
4. Two gear wheels mesh externally and are to give a velocity ratio of 3. The teeth are involute form with diametral pitch 2. Use standard addendum values and pressure angle is 18°. Determine 1. the number of teeth in each wheel to avoid interference, 2. the lengths of path and arc of contact, 3. the number of pairs of teeth in contact and 4. the angle of rotation of the pinion whilst any one pair of teeth are in contact.
5. State the essential requirements of the mating profiles of two gear teeth for a constant velocity ratio and show how these are satisfied by involute profiles. Determine graphically or otherwise, the length of path of contact when a pinion of 18 teeth meshes with an internally toothed wheel with 72 teeth. The pressure angle is 20° and the diametral pitch is equal to 2. The addenda on pinion and wheel are 0.9 and 0.4 cm respectively.
6. A pinion of 20 involute teeth and 12.5 cm pitch circle diameter drives a rack. The addendum of both the pinion and the rack is 6 mm. What is the least pressure angle which can be used to avoid undercutting? With this pressure angle find the length of arc of contact and the minimum number of teeth in contact at a time.
7. A gear wheel having 20 teeth of involute form of diametral pitch 2 and pressure angle 20° drives another wheel of the same dimensions. Determine the addendum of the wheel so that the arc of contact is maximum possible. What is the length of such an arc?
8. A gear wheel having 20 involute teeth of 1.25 cm circular pitch is to be generated by means of a straight rack cutter. The addendum of the cutter and of the wheel is 0.4 cm. What is the smallest pressure angle which may be employed, if undercutting is to be avoided? Calculate, from first principles, the length of the arc of contact when two such wheels, each of 20 teeth mesh together correctly.
9. Two equal gear wheels of 15 cm pitch circle diameter and diametral pitch 4 are in mesh. The teeth are of involute form with pressure angle 20°. Determine the

minimum addendum necessary if there are to be at least two pairs of teeth in contact.

10. A pinion with 24 involute teeth of 15 cm pitch circle diameter drives a rack. The addendum of the pinion and rack is 0.6 cm. What is the least pressure angle which can be used if interference is to be avoided? Using this pressure angle, find the length of arc of contact and the minimum number of teeth in contact at one time.

Chapter 7
Helical, Spiral, Worm and Bevel Gears

7.1 Involute Helicoid

Cut a paper into the shape of a parallelogram and wrap it around a cylinder. The edge of the paper makes an angle ψ_b called the base helix angle. When you unwrap the paper, keeping it tight, so that it is tangential to the cylinder, see Figure 7.1, you are generating several involute profiles placed side by side starting from a line inclined at the base helix angle ψ_b to the cylinder axis. The surface thus generated is an involute helicoid and forms the shape of a tooth on a helical gear. If the helix angle is clockwise, then we have a right-hand helical tooth, otherwise it is a left hand helical tooth. The plan view of a left-handed helical gear is shown in Figure 7.2.

To understand how two helical gears can mate, fold two A-4 size sheets to form roughly 30 cm by 5 cm size papers. On one of them draw lines representing right-hand teeth at any suitable helix angle and on the other draw lines with the same helix angle but with left-hand teeth. Fold them to form two cylinders and place them one above the other to understand how helical gears mate. Notice that these two gears have to be of opposite hand.

Because of the helix angle, there is always a thrust accompanied with the transmitted force. The net force is in the normal direction to the teeth in action as shown in Figure 7.2.

7.2 Helical Gear Tooth Relations

Figure 7.3 shows a helical rack in which lines *LM* and *NO* represent the center lines of two consecutive teeth.

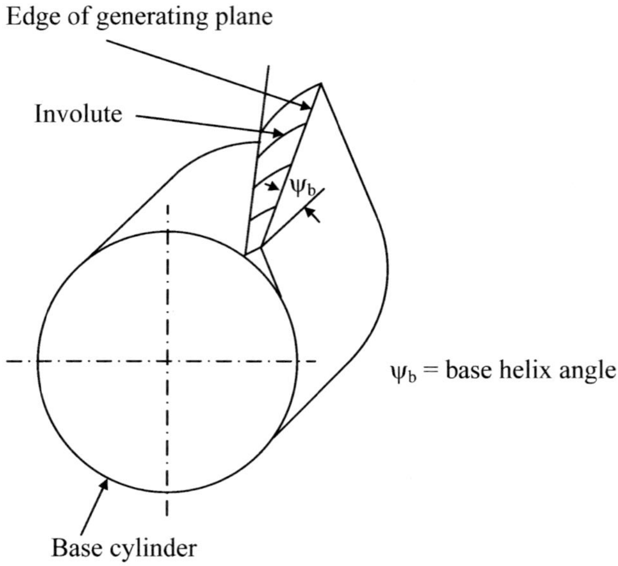

Fig. 7.1 Shape of a tooth on a helical gear

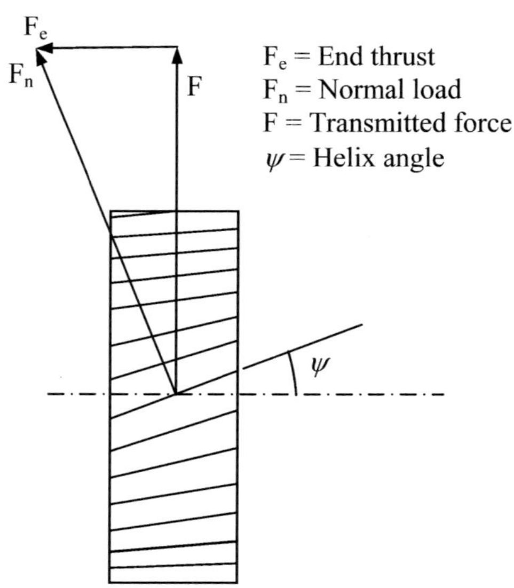

Fig. 7.2 Left-handed helical gear

Helix Angle, ψ

Angle made by a tooth center line with the gear axis.

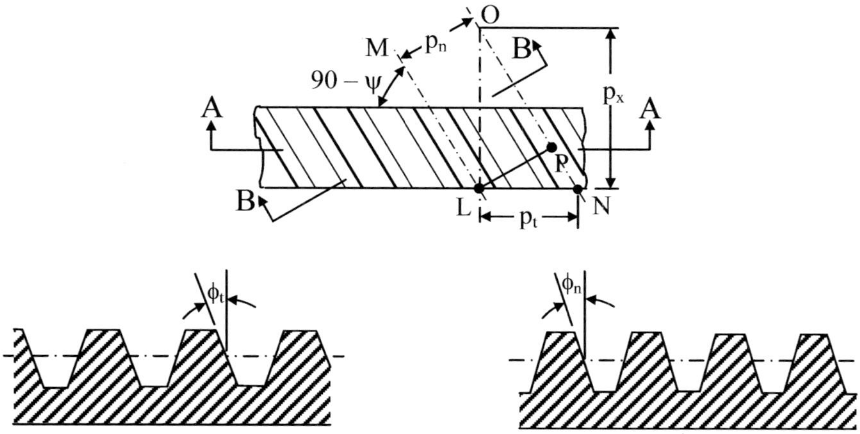

Fig. 7.3 Helical rack

Lead Angle, α

Complement of helix angle is $90 - \psi$.

Right-Handed Gear

Helical gear with helix angle in clockwise direction.

Left-Handed Gear

Helical gear with helix angle in anti-clockwise direction.

Transverse [Circular] Pitch, p_t

Distance between two consecutive teeth in the transverse direction.

Normal Pitch [Pitch], p_n

Distance between two consecutive teeth in the normal direction.

$$p_n = p_t \cos \psi \tag{7.1a}$$

Axial Pitch, p_x

Distance between two consecutive teeth in the axial direction.

$$p_x = \frac{p_t}{\tan \psi} \tag{7.1b}$$

Transverse [Circular] Diametral Pitch, p_{td}

$$p_{td} = \frac{\pi}{p_t} \tag{7.2a}$$

Normal Diametral Pitch [Diametral Pitch], p_{nd}

$$p_{nd} = \frac{\pi}{p_n} \tag{7.2b}$$

$$p_{nd} = \frac{p_{td}}{\cos \psi} \tag{7.3}$$

Transverse [Circular] Pressure Angle, φ_t

Pressure angle of helical rack tooth in transverse plane, see section A–A in Figure 7.3.

Center Distance for Helical Gear Pair, C

$$\begin{aligned} C &= \frac{t_2 + T_3}{2p_{td}} \\ &= \frac{t_2 + T_3}{2p_{nd} \cos \psi} \end{aligned} \tag{7.4}$$

Normal Pressure Angle, φ_n [Pressure Angle]

Pressure angle of helical rack tooth in normal plane, see section B–B in Figure 7.3.

$$\frac{\tan \varphi_n}{\tan \varphi_t} = \cos \psi \tag{7.5}$$

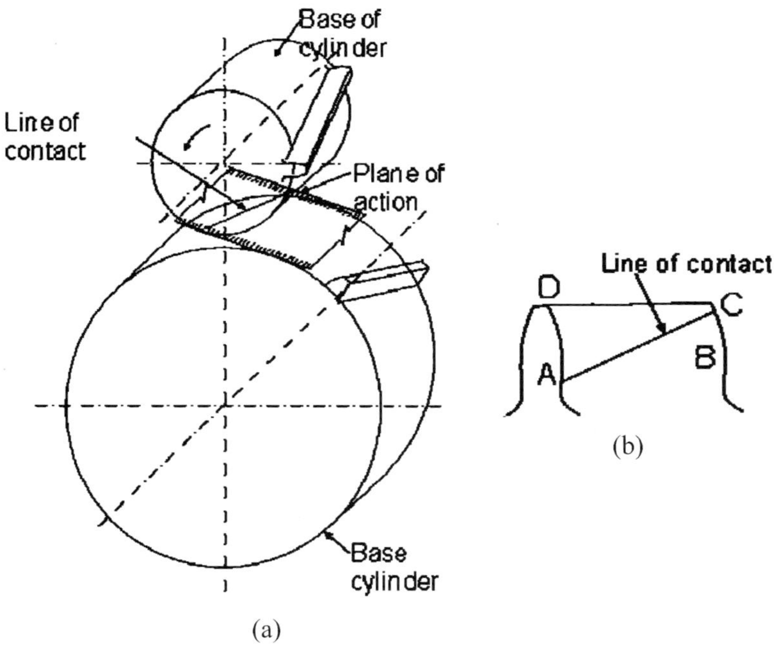

Fig. 7.4 (a) Two base cylinders with one helical tooth on each. (b) Contact pattern

Helical Gear Contact Ratio

Ratio of face width F to transverse pitch p_t. General recommendation for helical gears is

$$F > \frac{1.15 p_t}{\tan \psi} \tag{7.6}$$

7.3 Contact of Helical Gear Teeth

Figure 7.4a shows the two base cylinders each with one helical tooth. The top cylinder gear tooth is left-handed and the bottom one is right-handed. The plane of action is tangential to both the base cylinders. The contact begins at one end of the tooth, and if it is a driver, it starts at a point B on the flank as shown in Figure 7.4b. As the driver rotates, the contact gradually increases as shown, with the line of contact at an instant given by the line AC. The contact afterwards gradually decreases during the disengagement and finally disappears at the addendum circle point D on the face. If the gear is a driven gear, the contact begins at a point on the addendum circle and disappears somewhere in the bottom region of the flank.

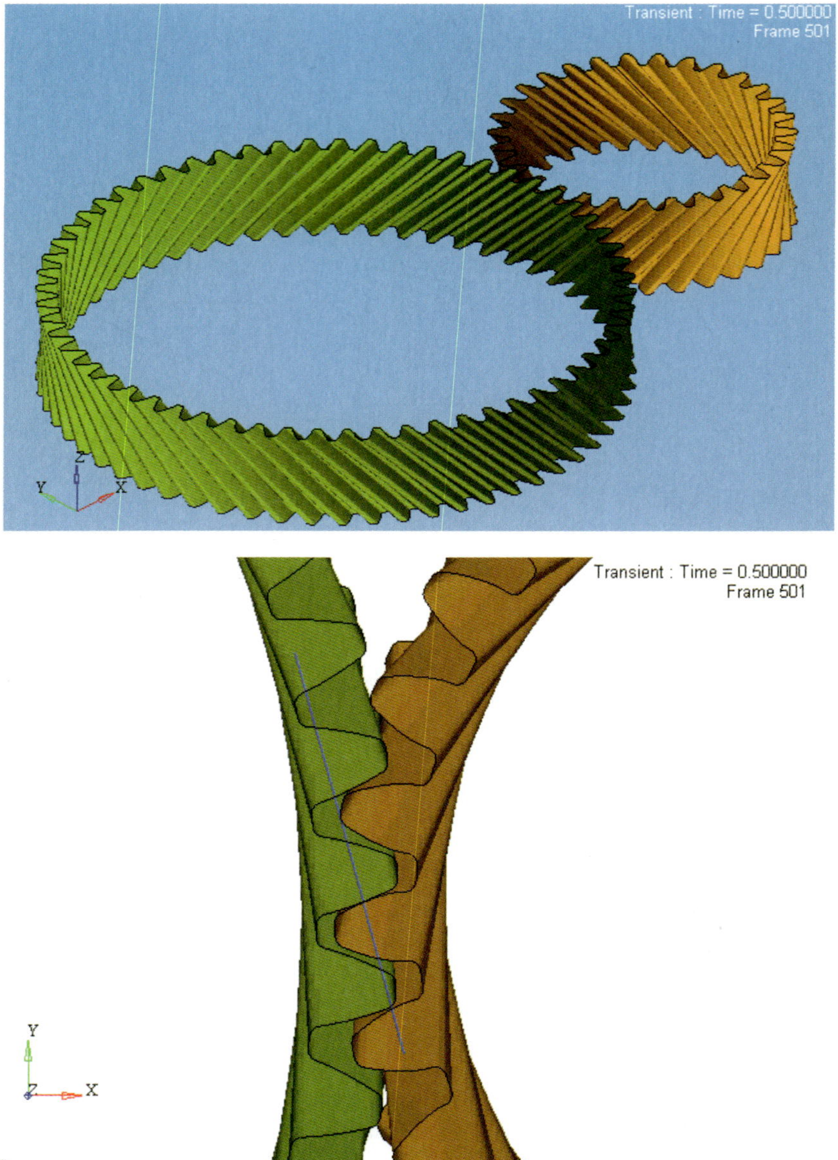

(c)

(d)

Fig. 7.4 (Continued) (c) Two helical gears in action. (d) Engagement of helical gears

In spur gears, the contact takes place on a line parallel to the axis abruptly and disengagement is also abrupt. In helical gears, the loading and unloading is gradual and therefore, the action is smoother and less noisy.

Two helical gears in action are shown in Figures 7.4c and d.

7.4 Helical Gear Calculations

A spur gear pair of 20° full depth with 24 and 60 teeth having a diametral pitch 6 is proposed to be replaced by a helical gear pair of the same pressure angle and diametral pitch. Keep the center distance to be the same as in a spur gear pair and the side thrust to be as small as possible.

1. For the existing spur gear pair, the center distance of the shafts is $C = (t_2 + T_3)/2p_d$.
2. With $t_2 = 24$, $T_3 = 60$ and $p_d = 6$, we have $C = 7$ cm.
3. Speed ratio $R = 60/24 = 2.5$.
4. For helical gear pair, the center distance is $C = (t_2+T_3)/2p_{td}$, see equation (7.4).
5. Using $T_3 = 2.5t_2$ and $C = 7$ cm, $p_{td} = 3.5t_2/14 = t_2/4$.
6. Since the teeth number should be an integer, let $t_2 = 20$, then $T_3 = 50$ and $p_{td} = 5$.
7. From equation (7.3),
$$p_{nd} = \frac{p_{td}}{\cos\psi}$$
Therefore $\cos\psi = 5/6 = 0.8333$.
8. Hence, the helix angle $\psi = 33°34'$.
9. The above helix angle is too large and hence the side thrust.
10. Let $t_2 = 22$, then $T_3 = 55$ and $p_{td} = 5.5$.
11. This gives
$$\psi = \cos^{-1}\frac{5.5}{6} = 23°38'$$
12. Selecting the above gears, from equation (7.2a), $p_t = \pi/5.5 = 0.571$.
13. Therefore the face width is, from equation (7.6),
$$F > \frac{1.15 \times 0.571}{\tan 23°38'} = 1.5 \text{ cm}$$

7.5 Spiral [Crossed Helical] Gears

The only condition to be satisfied here is that both the gears should have the same normal pitch. The contact takes place at only one point; therefore, their load carrying capacity is very small. Typical application example is the distributor drive in an automobile engine.

For two mating spiral gears, let p_n be the normal pitch; p_{t2} the transverse (circular) pitch of wheel 2; p_{t3} the transverse (circular) pitch of wheel 3; t_2 the number of teeth on wheel 2; T_3 the number of teeth on wheel 3; ψ_2 the helix angle of gear 2; ψ_3 the helix angle of gear 3; θ the shaft angle; D_2 the pitch circle diameter of wheel 2; and D_3 the pitch circle diameter of wheel 3.

Then, the following relations are valid:

$$R = \frac{T_3}{t_2}$$

$$p_{t2} = \frac{p_n}{\cos \psi_2}$$

$$p_{t3} = \frac{p_n}{\cos \psi_3}$$

$$D_2 = \frac{t_2 p_{t2}}{\pi} = \frac{t_2 p_n}{\pi \cos \psi_2} = \frac{t_2}{p_{nd} \cos \psi_2}$$

$$D_3 = \frac{T_3 p_{t3}}{\pi} = \frac{T_3 p_n}{\pi \cos \psi_3} = \frac{T_3}{p_{nd} \cos \psi_3} \tag{7.7}$$

The center distance between the shafts is then given by

$$C = \frac{1}{2}(D_2 + D_3) = \frac{t_2 p_n}{2\pi}\left(\frac{1}{\cos \psi_2} + \frac{R}{\cos \psi_3}\right)$$

$$= \frac{t_2}{2 p_{nd}}\left(\frac{1}{\cos \psi_2} + \frac{R}{\cos \psi_3}\right) \tag{7.8}$$

Note that the two mating crossed helical gears could be of the same hand or opposite hand. The shaft angle will depend on the helix angles and their hand. In Figure 7.5a, both the mating gears are left-handed, then, the shaft angle $\theta = \psi_2 + \psi_3$. If the crossed helical gears are of opposite hand, then as shown in Figure 7.5b, we have shaft angle $\theta = \psi_2 - \psi_3$. Hence the shaft angle for a spiral gear pair is

$$\theta = \psi_2 \pm \psi_3 \tag{7.9}$$

7.6 Worm Gearing

Generally used for transmitting power at very high velocity ratio in one stage between two non-intersecting shafts. The worm is shown in mesh with the gear wheel (also called worm wheel) in Figure 7.6. The pitch diameters of the worm and wheel are denoted D_w and D_g, and the helix angle as ψ. The lead angle is $\alpha = 90 - \psi$.

Conditions for Worm and Wheel at Right Angles

The necessary conditions are

1. Axial pitch of the worm p_{aw} should be equal to circular pitch p_{cg} of the gear wheel, see Figure 7.6.
2. Lead angle of the worm α should be equal to helix angle ψ_g of the gear.

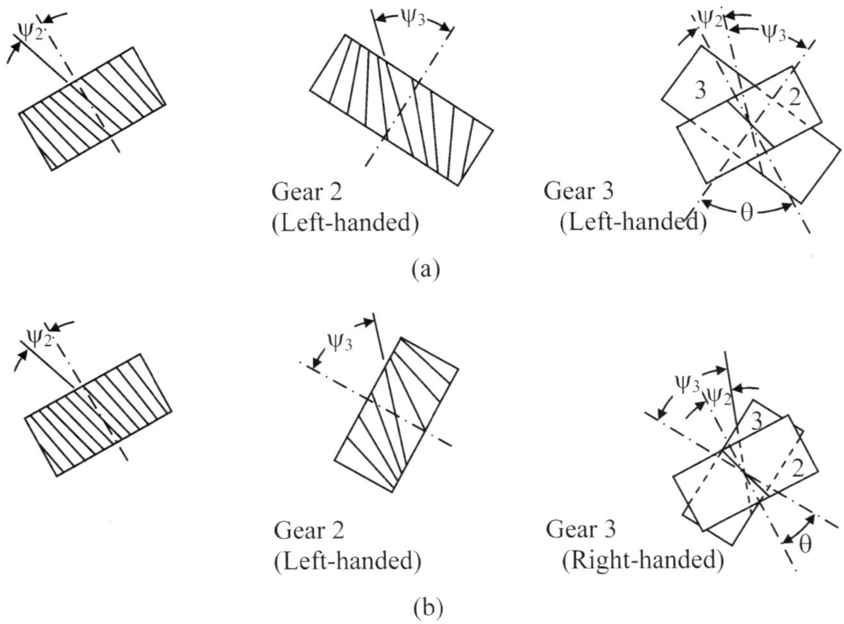

Fig. 7.5 (a) Two left-handed helical gears. (b) Two gears with opposite helix angles

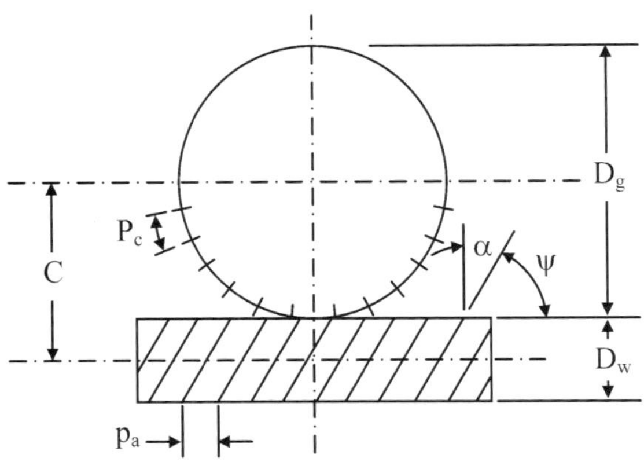

Fig. 7.6 Worm and worm gear

Lead of Worm

Distance traveled by a point on the worm thread in axial direction for one turn of the worm.

If one turn of the worm is unwrapped, it forms the hypotenuse of a right angled triangle as shown in Figure 7.7. If t_w is the number of threads on the worm, then the lead l is given by

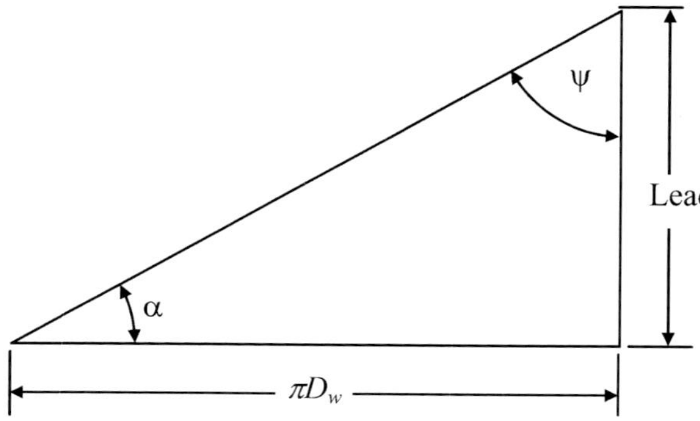

Fig. 7.7 One turn of the worm unwrapped

$$l = p_a t_w \tag{7.10}$$

From Figure 7.7,

$$\tan\alpha = \frac{p_a t_w}{\pi D_w} \tag{7.11}$$

If T_g is the number of teeth on the gear, its diameter can be obtained from

$$D_g = \frac{p_c T_g}{\pi} \tag{7.12}$$

The speed ratio is

$$\frac{n_w}{N_g} = \frac{T_g}{t_w} \tag{7.13}$$

Using (7.12) and (7.13), the speed ratio becomes

$$\frac{n_w}{N_g} = \frac{D_g}{D_w \tan\alpha} \tag{7.14}$$

For right-angled shafts, we have $\psi_w = 90 - \alpha$ and $\psi_g = \alpha$, therefore

$$\frac{n_w}{N_g} = \frac{D_g \cos\psi_g}{D_w \cos\psi_w}$$

$$= \frac{\pi D_g}{l} \tag{7.15}$$

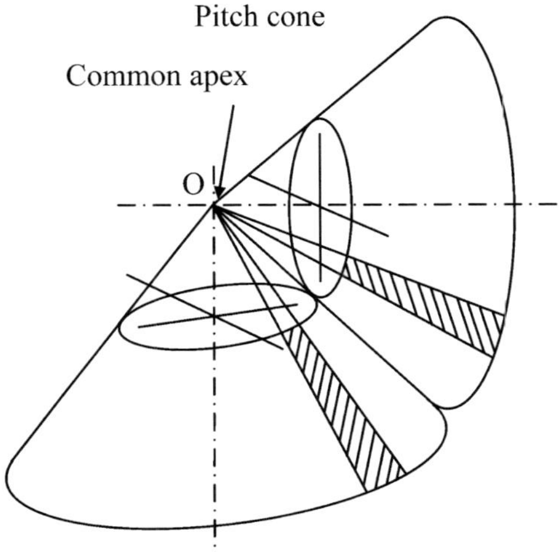

Fig. 7.8 Bevel gears

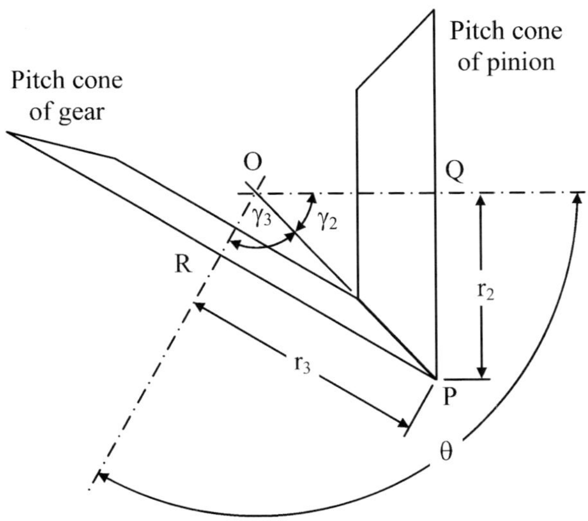

Fig. 7.9 Pitch cones of gear and pinion

7.7 Bevel Gears

The pitch cones roll over each other without slipping, with their apexes meeting at the same point as shown in Figure 7.8. Figure 7.9 shows two bevel gears in mesh with θ as their shaft angle. The shaft angle is the sum of the pitch cone angles.

$$\theta = \gamma_2 + \gamma_3 \tag{7.16}$$

The velocity ratio is

$$\frac{n_2}{N_3} = \frac{r_3}{r_2} = \frac{T_3}{t_2} \tag{7.17}$$

Since $OP = r_2/\sin\gamma_2 = r_3/\sin\gamma_3$,

$$\sin\gamma_2 = \frac{r_2}{r_3}\sin\gamma_3 \tag{7.18}$$

Using (7.16),

$$\sin\gamma_2 = \frac{r_2}{r_3}\sin(\theta - \gamma_2) \tag{7.19}$$

Expanding the above equation and dividing it by $\cos\gamma_2$, we get

$$\tan\gamma_2 = \frac{\sin\theta}{\cos\theta + \frac{r_3}{r_2}} \tag{7.20}$$

Similarly,

$$\tan\gamma_3 = \frac{\sin\theta}{\cos\theta + \frac{r_2}{r_3}} \tag{7.21}$$

For right angled shafts, $\theta = 90°$, therefore

$$\tan\gamma_2 = \frac{r_2}{r_3} \tag{7.22}$$

$$\tan\gamma_3 = \frac{r_3}{r_2} \tag{7.23}$$

7.8 Formation of Bevel Gears

Figure 7.10 illustrates the formation of a bevel gear tooth profile. The cone *OHI* is the base cone and *ODC* is the pitch cone. Imagine a paper wrapped around the cone with its edge along *OE*. When the paper is slowly unwrapped, say, to position *OG*, the edge *OE* generates a spherical involute *OF*. This spherical involute forms the profile of the bevel gear tooth. Since *OE* = *OF* at all stages, the point *E* moves on the surface of a sphere *HAI*.

It is difficult to layout the gear tooth in this form for a kinematic study; an approximation is made. The sphere *HAI* is replaced by a back cone *CBD* with back cone angle β, which is normal to the pitch cone *OCD*, i.e., $\beta = 90 - \gamma$. This back cone is also normal to the sphere *HAI* on the pitch circle *CD*. The teeth are laid out on the back cone developed surface as shown in the figure. This forms the tooth profile at the large end of the teeth, and all the tooth elements form straight lines with the pitch cone center O.

In the study of bevel gears, equivalent spur gears are used. The equivalent pitch radius of such a spur gear is

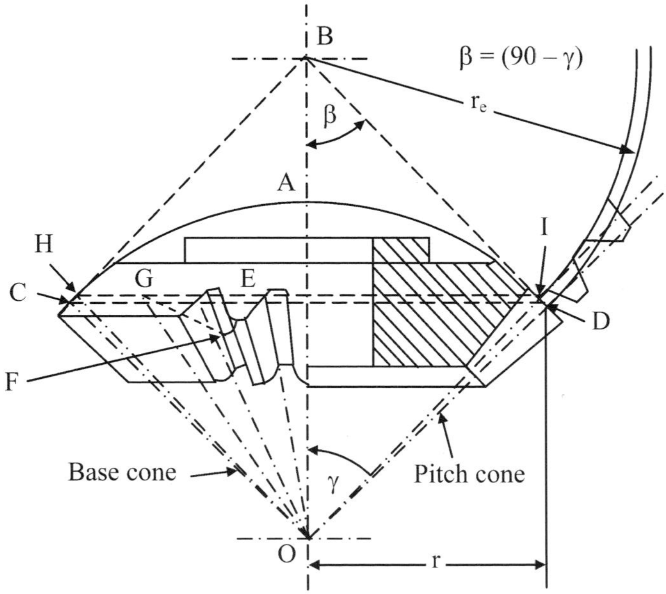

Fig. 7.10 Straight bevel gear tooth

$$r_e = \frac{r}{\sin \beta} = \frac{r}{\cos \gamma} \tag{7.24}$$

The number of teeth on the equivalent spur gear is

$$t_e = \frac{2\pi r_e}{p} \tag{7.25}$$

where p is the circular pitch of the bevel gear at the large end of the teeth. This is a formative number of teeth; therefore, it may not be a whole number.

Consider a 2-DP straight bevel pinion ($T_p = 14$ teeth) which drives a gear ($T_g = 24$ teeth). The shaft angle is 90° and the pressure angle is 20°. The following proportions are used to calculate various parameters of the bevel gear pair.

1. Working depth is $2/p_d = 1$ cm.
2. Whole depth is $(2.188/p_d) + 0.005 = 1.099$ cm.
3. Addendum of gear

$$a_G = \frac{0.54}{p_d} + \frac{0.4}{p_d}\left(\frac{T_p}{T_G}\right)^2 = 0.338 \text{ cm}$$

4. Addendum of pinion

$$a_p = \frac{2}{p_d} - a_G = 0.662 \text{ cm}$$

5. Dedendum of gear

$$d_g = \frac{2.188}{p_d} - a_G = 0.756 \text{ cm}$$

6. Dedendum of pinion

$$d_p = \frac{2.188}{p_d} - a_p = 0.432 \text{ cm}$$

7. Circular pitch $p_c = \pi / p_d = 1.571$ cm.
8. Circular thickness of gear

$$t_G = \frac{1}{2} p_c - (a_p - a_G) \tan \varphi = 0.6675 \text{ cm}$$

9. Circular thickness of pinion $t_p = p_c - t_g = 0.9035$ cm.
10. Pitch cone angle of pinion

$$\gamma_2 = \tan^{-1} \frac{r_2}{r_3} = \tan^{-1} \frac{14}{24} = 30.26°$$

11. Pitch cone angle of gear

$$\gamma_3 = \tan^{-1} \frac{r_3}{r_2} = \tan^{-1} \frac{24}{14} = 59.74°$$

12. Pitch radius of the equivalent spur pinion $r_2 = 14/(2 \times 2) = 3.5$ cm.
13. Pitch radius of the equivalent spur gear $r_3 = 24/(2 \times 2) = 6$ cm.
14. Base radius of the equivalent spur pinion $r_{e2} = 3.5/\cos 30.26 = 4.05$ cm.
15. Base radius of the equivalent $r_{e3} = 6/\cos 59.74 = 11.91$ cm.

7.9 Solved Problems

Solved Problem 7.1

Two shafts are to be connected by helical gears for a speed reduction of 1.5:1. The center distance between the shafts is 6 cm. Using 4-DP 20° full depth gears determine the number of teeth and pitch diameter of each gear and also the diametral pitch. The helix angle can be around 20°.

1. With helix angle 20°, the transverse pitch diameter $p_{td} = 4 \times \cos 20 = 3.7587$ cm.
2. In equation (7.4) $(t_2 + T_3)/2C = p_{td}$, substitute for $T_3 = 1.5t_2$ and other known values $2.5t_2/12 = 3.7587$. Therefore the number of teeth on the pinion and the gear are $t_2 = 18$ and $T_3 = 27$.
3. The transverse pitch diameter is

$$p_{td} = \frac{18+27}{2 \times 6} = 3.75 \text{ cm}$$

4. The helix angle for this pair is $\psi = \cos^{-1}(3.75/4) = 20.364°$.
5. Pitch diameters are $D = 18/3.75 = 4.8$ cm and cm $D_3 = 27/3.75 = 7.2$ cm.
6. Diametral pitch $p_{nd} = 3.75/\cos 20.364 = 4$.

Solved Problem 7.2

It is proposed to replace a 6-DP 20° full depth spur gear pair of 32 and 80 teeth by a helical gear pair, without altering the center distance and speed ratio. Keeping the helix angle as small as possible, design the gear pair keeping the width to be below 1.25 cm.

1. For spur gear pair, the center distance is $C = (32+80)/(2 \times 6) = 9.333$.
2. The speed ratio is $R = 80/32 = 2.5$.
3. For the helical gear pair $T_3 = 2.5t_2$. From equation (7.4)

$$p_{td} = \frac{3.5t_2}{2 \times 9.3333} = 0.185t_2$$

4. From the face width condition in equation (7.6) $1.25 > 1.15\pi/p_{td} \tan\psi$. Or $1.25 > 1.15\pi/p_{nd} \sin\psi$.
5. Therefore

$$\psi > \sin^{-1}\frac{1.15\pi}{6 \times 1.25} = 28.8°$$

6. Then $p_{td} = 6\cos 28.8 = 5.25784$.
7. Number of teeth on pinion $t_2 = 5.25784/0.1875 = 28$.
8. Number of teeth on gear is 70.
9. Work backwards to get $p_{td} = 0.1875 \times 28 = 5.25$ cm.
10. $\psi = \cos^{-1}(5.25/6) = 28.95°$.
11. Transverse circular pitch $p_t = \pi/5.25 = 0.5984$.
12. Face width $F = (1.15 \times 0.5984)/\tan 28.95 = 1.225$ cm.

Solved Problem 7.3

A 5-DP 24 tooth spur pinion drives two gears, one of 36 teeth and the other 80. It is required to replace all three gears by helical ones without altering the center distances, diametral pitch as well as gear ratios. Use 20° full depth involute system.

1. For spur gears, the center distances are $C_1 = (24+36)/(2 \times 5) = 6$ cm and $C_2 = (36+80)/(2 \times 5) = 11.6$ cm.
2. The speed ratios are $R_1 = 36/24 = 1.5$ and $R_2 = 80/36 = 2.222$.
3. For the helical gear pair 1: $p_{td1} = 2.5t_2/(2 \times 6) = 0.20833t_2$.
4. For the helical gear pair 2: $p_{td2} = 3.222t_2/(2 \times 11.6) = 0.1388t_2$.

5. To keep $p_{nd} = 5$ and $p_{td} < 5$, we try different combinations, and finally p_{td} is chosen as 3.75.
6. This gives the following for gear pair 1: $t_2 = 18$ and $T_3 = 27$.
7. For gear pair 2: Since there is a common gear, gear 2 in this pair should have same teeth as gear 3 of the gear pair 1, i.e., $t_2 = 27$ and $T_3 = 60$.
8. So, $\psi = \cos^{-1}(3.75/5) = 41.41°$.
9. Face width from equation (7.6)

$$F = \frac{1.15\pi}{3.75\tan 41.41} = 1.1 \text{ cm}$$

10. Gear diameters of pair 1: $D_2 = 18/3.75 = 4.8$ cm and $D_3 = 27/3.75 = 7.2$ cm.
11. Gear pair 2: $D_2 = 27/3.75 = 7.2$ cm and $D_3 = 60/3.75 = 16$ cm.

Solved Problem 7.4

Using a 30 teeth 30° helix angle and 2-DP pinion, two shafts are to be connected by crossed helical gears to give an angular velocity ratio 1.5. The center distance between the shafts is 20 cm. Determine the shaft angle using a gear having the same hand as the pinion.

1. Using the second relation in equation (7.7), $p_{t2} = p_n/\cos\psi_2$, we have $30/D_2\cos 30 = 2$. Hence the diameter of pinion $D_2 = 17.3205$ cm.
2. Since the center distance is 20, $D_3 = 40 - 17.3205 = 22.6795$ cm.
3. Number of teeth on gear 3 is $T_3 = 30 \times 1.5 = 45$.
4. Therefore $45/22.6795\cos\psi_3 = 2$, i.e., $\psi_3 = 7.2134°$.
5. Since both the gears are of same hand, the shaft angle $\theta = \psi_2 + \psi_3 = 37.2134°$.

Solved Problem 7.5

Design a proper spiral gear pair (or a helical one) to connect two shafts whose center distance is 12.5 cm. The angular velocity ratio is to be 2.0.

1. Choose gears of 4-DP (or any value you like).
2. From equation (7.4)

$$12.5 = \frac{t_2 + 2t_2}{2 \times 4\cos\psi_2}$$

3. This gives $t_2/\cos\psi_2 = 33.333$.
4. Choose $t_2 = 25$.
5. Choose a left-handed helical pinion with a helix angle

$$\psi_2 = \cos^{-1}\frac{25}{33.333} = 41.41°$$

6. $T_3 = 50$.

7. For a helical gear pair, the two shafts are non-intersecting and the gears should be opposite in hand. In that case, the helix angle of both the gears should be same. Therefore, choose a right-handed gear with helix angle 41.41°.
8. You may like to try another pair with a suitable shaft angle and find the gear helix angle.

Solved Problem 7.6

The pinion of a spiral gear pair has 30 teeth and a normal diametral pitch 3. The center distance between the shafts is 25 cm; the shaft angle is 60° and the velocity ratio 3.5. Calculate the helix angles and pitch diameters if the gears have the same hand.

1. From equation (7.8) the center distance

$$C = \frac{t_2 p_n}{2\pi}\left(\frac{1}{\cos\psi_2} + \frac{R}{\cos\psi_3}\right)$$

2. Substituting the known values

$$25 = \frac{30\pi}{3 \times 2\pi}\left(\frac{1}{\cos\psi_2} + \frac{3.5}{\cos(60 - \psi_2)}\right)$$

3. Therefore

$$\left(\frac{1}{\cos\psi_2} + \frac{3.5}{\cos(60 - \psi_2)}\right) = 5$$

   or $\psi_2 = 41°$ and $\psi_3 = 19°$.
4. From equation (7.7)

$$D_2 = \frac{t_2}{p_{nd}\cos\psi_2} = \frac{30}{3\cos 41} = 13.25 \text{ cm}$$

$$D_3 = \frac{T_3}{p_{nd}\cos\psi_3} = \frac{30 \times 3.5}{3\cos 19} = 37.02 \text{ cm}$$

Solved Problem 7.7

A double threaded worm which has a lead of 5 cm drives a worm gear. The velocity ratio is 20 and the shaft angle is 90°. Determine the pitch diameter of worm and worm gear, if the center distance is 20 cm.

1. From equation (7.15)

$$R = \frac{n_w}{N_g} = \frac{\pi D_g}{l}$$

2. Therefore gear diameter can be obtained as $20 = \pi D_g/l$ and we get $D_g = 100/\pi = 31.831$ cm.
3. Since $D_g + D_w = 40$, the worm diameter is $D_w = 8.169$ cm.

Solved Problem 7.8

A right-angled worm and worm gear set is required to have a velocity ratio 25. Using a center distance of 15 cm and axial pitch 1 cm of the worm, determine the smallest diameter worm that can be used.

1. Using equation (7.15),

$$25 = \frac{\pi D_g}{l} = \frac{\pi(30 - D_w)}{l}$$

2. Therefore $D_w = 30 - 7.96l$.
3. From equation (7.10) $l = p_a t_w = t_w$.
4. The worm diameter will be smallest when it has a maximum number of threads, usually not more than three. Therefore choosing $t_w = 3$, we have $D_w = 6.12$ cm.

Solved Problem 7.9

A four threaded worm drives a 48 tooth worm gear having a pitch diameter 18 cm and helix angle 20°. Calculate the lead and pitch diameter of the worm if the shaft angle is 90°.

1. $\psi_g = 20°$ and since the shafts are at right angles, $\psi_w = 70°$.
2. Using equation (7.15)

$$\frac{n_w}{N_g} = \frac{D_g \cos \psi_g}{D_w \cos \psi_w}$$

we have

$$\frac{48}{4} = \frac{18 \cos 20}{D_w \cos 70}$$

3. Therefore the worm diameter is $D_w = 4.12$ cm.
4. Once again using equation (7.15), $n_w/N_g = \pi D_g/l$, we get $18\pi/l = 12$.
5. The lead is therefore $l = 4.712$ cm.

Solved Problem 7.10

A pair of straight tooth bevel gears is to be used for a shaft angle of 90°. If the driver is to have 20 teeth and speed ratio 3.5, determine the pitch angles.

1. From equation (7.22) $\tan \gamma_2 = r_2/r_3 = 3.5$.
2. Therefore $\gamma_2 = 74.05°$.

3. For the driven gear, $\tan\gamma_3 = 1/3.5$ and $\gamma_3 = 15.95°$.
4. Or $\gamma_3 = 90 - 74.05° = 15.95°$.

Solved Problem 7.11

Two straight tooth bevel gears having 16 and 24 teeth are to be mounted at a shaft angle 120°. What are the pitch angles?

1. From equation (7.20),

$$\tan\gamma_2 = \frac{\sin\theta}{\cos\theta + \frac{r_3}{r_2}} = \frac{\sin 120}{\cos 120 + 1.5}$$

2. Therefore the pinion pitch angle

$$\gamma_2 = \tan^{-1}\frac{0.866}{1} = 40.89°$$

3. Similarly from (7.21), the driven member pitch angle is

$$\gamma_3 = \tan^{-1}\frac{0.866}{0.166} = 79.1°$$

7.10 Additional Problems

1. Two spiral gear wheels A and B of the same hand and equal diameter are used in a machine tool drive. The normal pitch is 1.25 cm. The center distance should be around 15 cm and the shaft angle is 80°. The speed of shaft A is 1.25 times that of shaft B. Determine (a) the spiral angle of each wheel and (b) the number of teeth on each wheel.
2. A pair of spiral gears is required to connect two shafts 17.5 cm apart, the shaft angle being 70°. The speed ratio is 1.5 and the faster wheel has 80 teeth and a pitch circle diameter of 10 cm. Find the spiral angles of each wheel.
3. Two shafts are to be connected by spiral gears with a velocity ratio 3:1. The shaft angle is 45° and the least distance between the shaft axes is to be 22.5 cm. The normal diametral pitch is to be 5 and the pinion is to have 20 teeth. Determine the pitch circle diameters and the spiral angles if they are of the same hand.
4. Two spiral gear wheels A and B have 45 and 15 teeth at spiral angles 20° and 50° respectively. Both wheels are of the same hand and wheel A is 15 cm diameter. Find the distance between the shafts and the angle between the shafts.
5. Design a worm wheel to be driven by a double threaded worm with a lead of 4 cm. The velocity ratio required is 16 and the shaft angle is to be 90°. The center distance can be around 25 cm.

6. A worm drives a 48 tooth worm gear having a pitch diameter 18 cm and helix angle 20°. Design the worm to obtain a speed ratio of 12. Use a shaft angle 90°.
7. Design the rolling cones for a 3:1 ratio and a 60° included angle between the shafts and make a layout.
8. A bevel gear pair is required to transmit motion with a velocity ratio 4:1. The shaft angle is 40°. Design the rolling cones.

Chapter 8
Gear Trains

8.1 Classification of Gear Trains

Simple Gear Train

Gear train which has no more than one gear mounted on a shaft.

Compound Gear Train

Two or more gear pairs arranged in series with at least one shaft carrying more than one gear.

Reverted Gear Train

A compound gear train with the driving and driven gears on coincident axes.

Epicyclic [Planetary] Gear Train

One gear in the train rotates over another gear causing an epicyclic path.

Differential Gear Train

An epicyclic gear train when all the members are free to rotate.

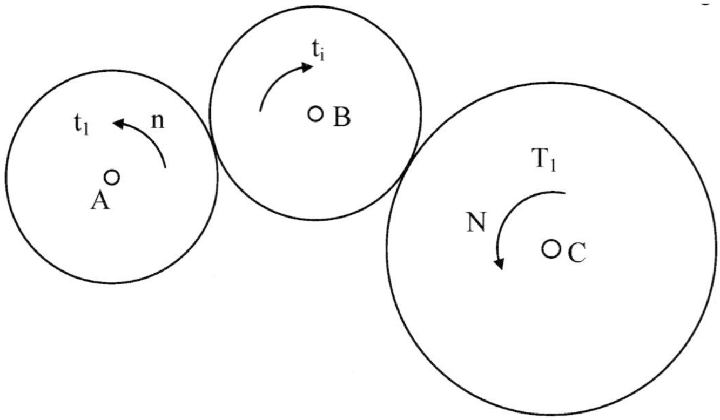

Fig. 8.1 Simple gear train

Coupled Epicyclic Gear Train

Two or more epicyclic trains are coupled together.

8.2 Simple Gear Trains

Figure 8.1 shows a simple gear train. A is the driver, B is the idler and C is the driven member. The speed of the driver is represented by n and the driven member speed is N. For gears mounted on shafts A and B, the speed ratio is

$$\frac{n}{\omega_i} = \frac{t_i}{t_1} \tag{8.1}$$

where ω_i is the speed of the idler.

For gears mounted on shafts B and C,

$$\frac{\omega_i}{N} = \frac{T_1}{t_i} \tag{8.2}$$

Therefore,

$$R = \frac{n}{N} = \frac{T_1}{t_1} \tag{8.3}$$

and the speed ratio R is independent of the idler wheel. In fact there can be any number of idler wheels; the speed ratio is still governed by equation (8.3). The purpose of idler wheels is only to reverse the direction of speed or bridge a gap.

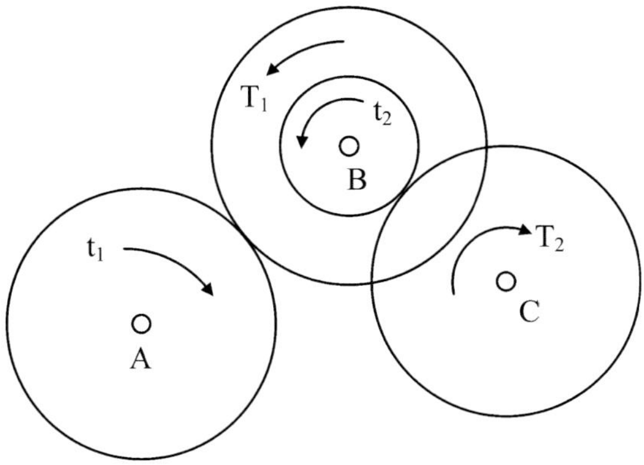

Fig. 8.2 Compound gear train

8.3 Compound Gear Trains

Figure 8.2 shows a compound train with two gear pairs t_1, T_1 and t_2, T_2. The speed of the wheels mounted on shaft B is the same, i.e., the follower of the first pair and the driver of the second pair have the same speed. Then

$$N_1 = n_1 \frac{t_1}{T_1} \tag{8.4}$$

$$n_2 = N_1 \tag{8.5}$$

$$N_2 = n_2 \frac{t_2}{T_2} \tag{8.6}$$

Therefore,

$$N_2 = n_1 \frac{t_1}{T_1} \frac{t_2}{T_2} \tag{8.7}$$

Hence the overall gear ratio is

$$R_{1,2} = \frac{n_1}{N_2} = \frac{T_1 T_2}{t_1 t_2} \tag{8.8}$$

In general for a compound train with n simple gear pairs,

$$R_{1,2,\ldots n} = \frac{T_1 T_2 \cdots T_n}{t_1 t_2 \cdots t_n} \tag{8.9}$$

Since the number of teeth is a whole number we can write

$$R = \frac{U}{u}$$
$$U = T_1 T_2 \ldots T_n$$
$$u = t_1 t_2 \ldots t_n \tag{8.10}$$

8.4 Synthesis of Gear Trains

Simple Gear Train

If the gear ratio desired can be expressed by two simple whole numbers U and u, and if these two whole numbers can represent convenient tooth numbers, then a simple gear train is possible.

As an example design a simple gear train for $R = 2.4$. We can write

$$R = \frac{U}{u} = \frac{24}{10} = \frac{12}{5}$$

Then we can choose gear trains with teeth 24 and 10 (if this is not good as interference may take place with the pinion having only 10 tooth) or 36 and 15, or 48 and 20, etc.

Compound Gear Train

When a simple train is not possible as above, a compound train may be tried.

Let us design a gear train for a speed ratio of 24.149. Here

$$R = \frac{U}{u} = \frac{24149}{1000}$$

We can factorize the above to give

$$R = \frac{U}{u} = \frac{19 \times 31 \times 41}{2 \times 5 \times 2 \times 5 \times 2 \times 5}$$

The above terms can be rewritten to represent convenient tooth numbers, e.g.,

$$R = \frac{U}{u} = \frac{76}{20} \times \frac{93}{24} \times \frac{41}{25}$$

Therefore, we can use a compound gear train with three simple gear trains, 76/20, 93/24 and 41/25 in series. Gear with 20 teeth in the first pair and the gear with 93 teeth in the second pair are mounted on the same shaft, shaft 2; Gear with 24 teeth in the second pair and the gear with 41 teeth in the third pair are mounted on the

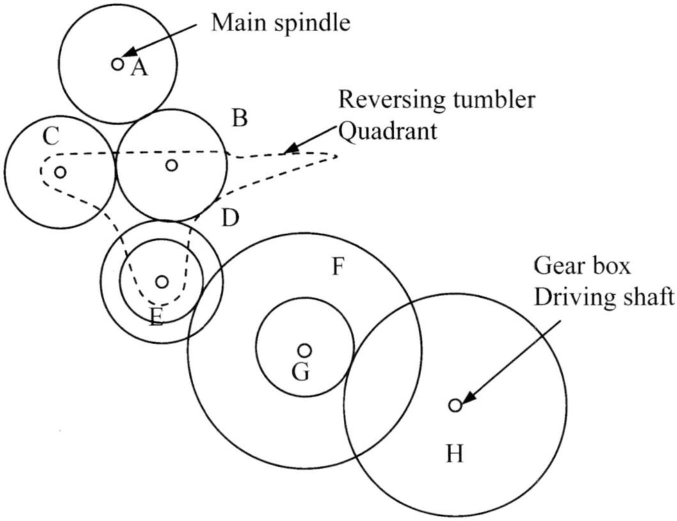

Fig. 8.3 Change gear train

same shaft, shaft 3. Shaft 1 with gear 76 teeth is the input shaft and shaft 4 with gear 25 teeth will form the driven shaft.

Note that it is not desirable to have a common factor between the teeth numbers in a pair, to distribute any successive impulsive forces. If there is a common factor, it is advisable to provide an additional tooth called hunting tooth.

8.5 Gear Train Applications to Machine Tools

Machine tools are classical examples of gear trains. They require many different drives, e.g., main drive to the spindle, automatic traverse of the cross slide, etc.

Change Gear Trains

Figure 8.3 shows a change gear train, usually employed for screw cutting in ordinary lathes. Gear A is mounted on the main spindle and it drives the output gear D through either one idler wheel B or set of idlers C and B. A reversing tumbler quadrant is used for the purpose of changing the direction of rotation of wheel D. This enables cutting of right-hand as well as left-hand screw threads on a lathe.

From gear D, the motion is transmitted to the quick change gear box, through a compound train. Pick off gears for the set of gear wheels E and F are provided to get the desired speed of the lead screw.

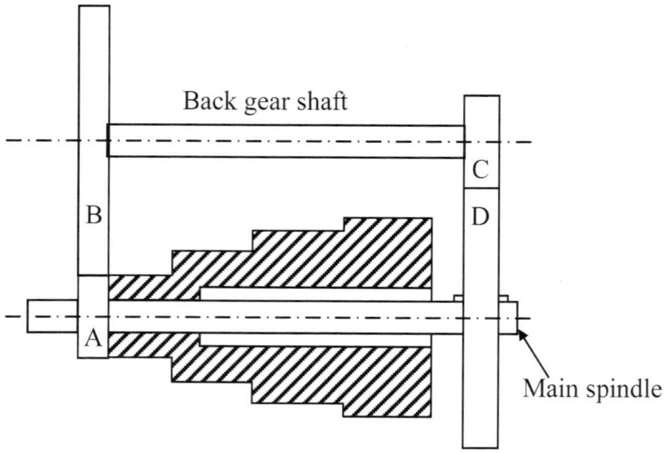

Fig. 8.4 Reverted gear train

Reverted Gear Train

The back gear arrangement in the head stock of a lathe is a good example of a reverted train. As shown in Figure 8.4, the cone pulley and gear A form an integral block and it may run loose or fast on the main spindle. Gears B and C are mounted on the back gear shaft, and the motion is transmitted through them to gear D, which is keyed to the main spindle.

All-Geared Head Stocks

Machine tool spindle drives need a number of speeds to achieve a minimum loss of cutting speed for any diameter of the job. It is found that the best arrangement for such a condition is to arrange the speeds in a geometrical progression.

Consider a six-speed gear box ($n_1, n_2, \ldots, n_6$) with three shafts. The speeds are to be arranged in a geometrical progression, therefore

$$\frac{n_2}{n_1} = \frac{n_3}{n_2} = \frac{n_4}{n_3} = \cdots \frac{n_6}{n_5} = \varphi \tag{8.11}$$

Or

$$\log n_2 - \log n_1 = \log n_3 - \log n_2 = \log n_4 - \log n_3 = \cdots = \log \varphi \tag{8.12}$$

Using equation (8.12) a kinematic layout diagram may be drawn as shown in Figure 8.5a. Shaft A is the input shaft running at a constant speed. Shaft B is an intermediate shaft, which can be made to run at any of the three speeds indicated in the figure by shifting a gear block (Note that a shifting gear block can give a maximum

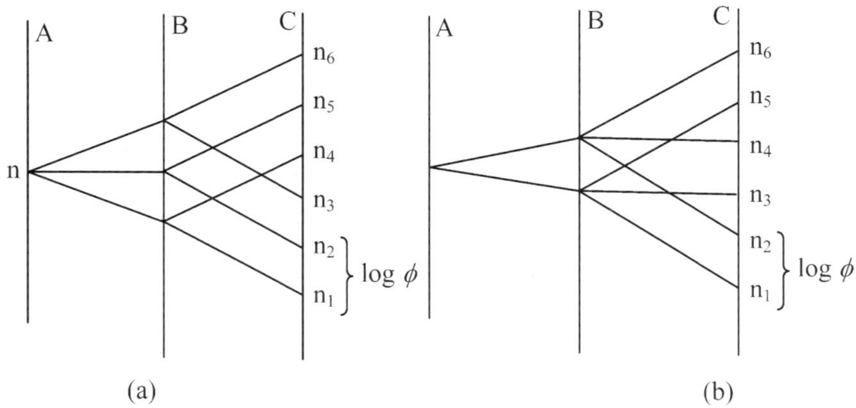

Fig. 8.5 Kinematic layout of a six-speed gear box

three speeds). The output shaft C can be made to run at six speeds, two different speeds for each speed of shaft B.

The arrangement can also be chosen as in Figure 8.5b, in which the intermediate shaft has only two speeds and corresponding to each of these speeds, the output shaft has three speeds. It is advisable to keep the input and intermediate shafts at as high speeds as possible so that the torque transmitted by them is restricted to a minimum value.

A Headstock of a machine tool is to have six speeds in the range 50 to 1600 RPM. The main drive runs at 1450 RPM. Suggest a kinematic layout and show schematically the arrangement of such a gear box.

1. Since there are six speeds, the value of φ from equation (8.11) can be determined as $\varphi^5 = 1600/50 = 32$ or $\varphi = 2$.
   It is necessary to have at least two reduction stages to obtain all the speeds on the output shaft. The input shaft can be coupled directly to the motor, in which case it will have a speed of 1450 RPM. This is shown in the ray diagram of Figure 8.6, where all the shaft speeds are marked on the log scale, with each division equal to $\log \varphi$.
2. The intermediate shaft is chosen to have only two speeds, so that its minimum speed can be kept as high as possible. These shaft speeds are chosen at 1600 and 200 RPM to be consistent with the ratios of speed reduction in the second stage to cover all the desired speeds on the output shaft. Note that the reductions from 1600 and 200 RPM on the intermediate shaft will be in the same ratio, i.e., there are three pairs of lines which are parallel on the diagram between the shafts B and C.
   The gear box layout to achieve the ray diagram thus drawn is shown in Figure 8.7. Note the locations of shifting gear blocks on input and output shafts and that the intermediate shaft carries only fixed gears. In the figure shown, gear pair 1 and 3 are in action. You can keep gear pair 1 in action and choose gear pairs 3 or 4

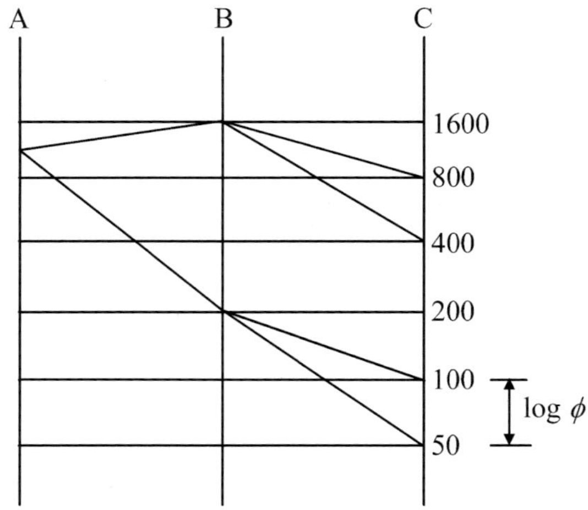

Fig. 8.6 Ray diagram for the example problem

or 5 to mate by shifting the gear block to appropriate position which gives the speeds 1600, 800 and 400 RPM to the output shaft. If you choose gear pair 2 to be in action, then speeds 200, 100 or 50 RPM are obtained. The tooth numbers can now be calculated as follows.

3. The speed ratio for gear set 1 is

$$\frac{n_1}{N_1} = \frac{T_1}{t_1} = \frac{1450}{1600} = \frac{29}{32}$$

   Let this be represented as $n_1/N_1 = 29x/32x$, where x is a common factor to be determined later.

4. The speed ratio for gear set 2 is

$$\frac{n_2}{N_2} = \frac{T_2}{t_2} = \frac{1450}{200} = \frac{29}{4}$$

   Let this be represented by $n_2/N_2 = 29y/4y$.

5. Since all the gears used are of the same diametral pitch, $t_1 + T_1 = t_2 + T_2$.
6. Therefore, $(29 + 32)x = (29 + 4)y$, i.e., $y = 1.85x$.
7. Since the tooth numbers are whole numbers, it is clear that the desired speeds of shaft B cannot be obtained as in the ray diagram.
8. One recommended combination of teeth of gear pairs 1 and 2 is $t_1 = 96$ and $T_1 = 87$ and $t_2 = 22$ and $T_2 = 161$.
9. Obviously, there is no unique solution. You may like to try finding the numbers of teeth for gear pairs 3, 4 and 5.

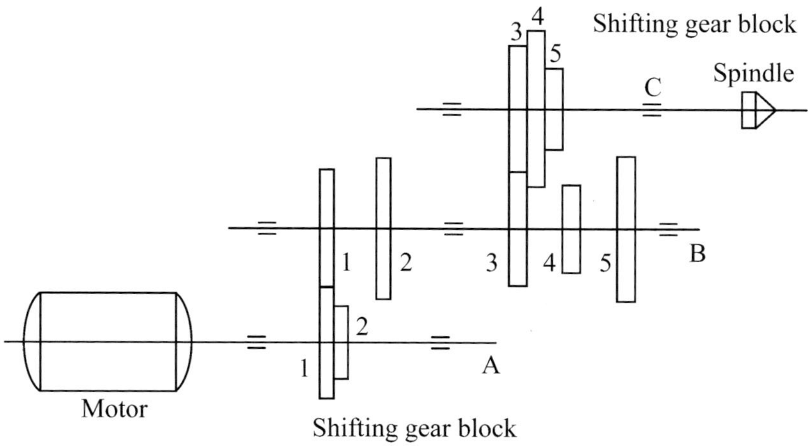

Fig. 8.7 Gear box layout for the example problem of Figure 8.6

8.6 Epicyclic Trains

Figure 8.8a shows the simplest epicyclic train in which shafts A and B represent the input and output members. These two shafts are connected by a simple reverted train. The block of gears having teeth T_1 and t_2 are mounted in a casing C. If C is fixed, we have a simple reverted train.

In epicyclic trains, the casing C is made free and in order to have one degree of freedom, either shaft A or shaft B is restrained from rotation. Let the shaft A be fixed, then the gear block along with the casing C rotates over the fixed gear on shaft A. Therefore, gear A becomes the sun wheel and the gear block becomes planet wheels and C is the planet carrier.

When the casing C is fixed, the overall speed ratio of the reverted train R_0 is defined as the basic ratio of the epicyclic train.

$$R_0 = -\frac{T_1}{t_1} \tag{8.13}$$

For the train in Figure 8.8a,

$$R_0 = \frac{T_1 T_2}{t_1 t_2} \tag{8.13a}$$

Other versions of the simple epicyclic train are given in Figures 8.8b and c. In the train of Figure 8.8b, an annulus is used on shaft B instead of an external gear of Figure 8.8a. Then, the basic ratio of the train is

$$R_0 = -\frac{T_1 T_2}{t_1 t_2} \tag{8.13b}$$

Table 8.1 Case A/C speed ratio calculations

Operation	Revolutions of A	B	C
Member C fixed	R_0	1	0
En bloc rotation	−1	− 1	−1
Net resulting motion	$R_0 - 1$	0	−1

The train in Figure 8.8b can be simplified to have one planetary wheel instead of a block, then the basic ratio becomes

$$R_0 = -\frac{T_1}{t_1} \tag{8.13c}$$

In all three cases above, we find that there are three common elements

1. Input shaft, A.
2. Output shaft, B.
3. Planet carrier, C.

In fact all epicyclic trains have these three elements; you must learn to recognize them in a given example. When C is fixed, the epicyclic train becomes a simple reverted train and the speed ratio of this train is a basic ratio. Three different arrangements of an epicyclic train are shown in Figure 8.9a (B fixed), Figure 8.9b (C fixed – reverted train) and Figure 8.9c (A fixed). Symbolically, all that you need is to write the train as shown in Figure 8.9d.

8.7 Inversions of Epicyclic Trains

Case A/C (B Fixed) Figure 8.8a

First consider when C is fixed as in a reverted train, rotate the shaft A, by sufficient number of rotations, to cause one complete revolution of shaft B. This requires R_0 revolutions of shaft A.

Next rotate the whole epicyclic train, en bloc by a -1 revolution, to bring the shaft B to its original position, during which the shaft A as well as casing C also have made a -1 revolution. These two processes are given in Table 8.1.

The speed ratio R_{AC} is therefore

$$R_{AC} = \frac{N_A}{N_C} = 1 - R_0 \tag{8.14a}$$

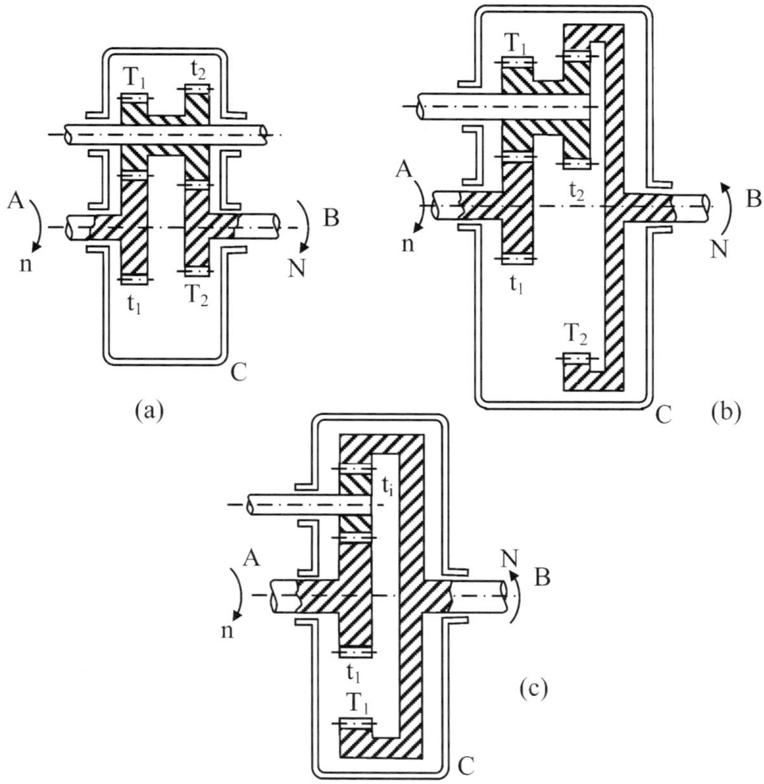

Fig. 8.8 (a) Single epicyclic train. (b) Output shaft with an internal gear. (c) Epicyclic train with a simple planet wheel

Table 8.2 Case B/C speed ratio calculations

Operation	Revolutions of		
	A	B	C
Member C fixed	R_0	1	0
En bloc rotation	$-R_0$	$-R_0$	$-R_0$
Net resulting motion	0	$1 - R_0$	$-R_0$

Case B/C (A Fixed) Figure 8.8c

First consider when C is fixed as in a reverted train, rotate the shaft A, by a sufficient number of rotations, to cause one complete revolution of shaft B. This requires R_0 revolutions of shaft A. (This operation is the same as above.)

Next rotate the whole epicyclic train, en bloc by $-R_0$ revolution, to bring the shaft A to its original position, during which the shaft B as well as casing C also have made $-R_0$ revolution. These two processes are given in Table 8.2.

The speed ratio R_{BC} is therefore

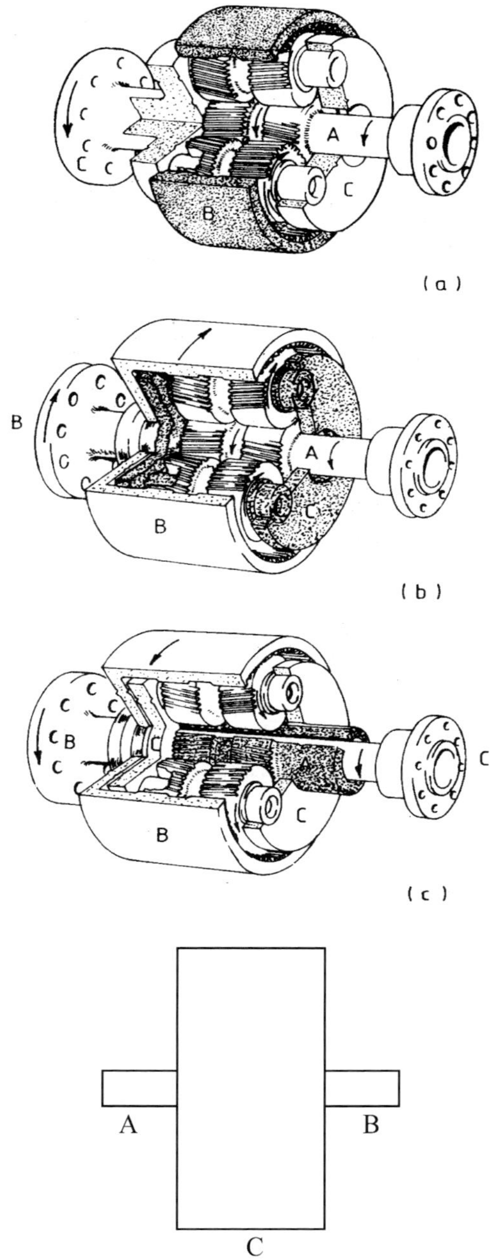

Fig. 8.9 (a), (b), (c) Different arrangements of epicyclic trains. (d) Elements in an epicyclic train

$$R_{BC} = \frac{N_B}{N_C} = \frac{R_0 - 1}{R_0} \tag{8.14b}$$

The main advantage of an epicyclic train is that a wide range of speed ratios can be obtained by simple alterations, see the example below.

The basic ratio of an epicyclic gear train is 1.05. Find the speed ratios that can be obtained by two epicyclic arrangements. What would be the speed ratios of the same trains, if the basic ratio is changed to -1.05?

- *Basic Ratio is* 1.05
  Case A/C (B Fixed): From (8.13)

$$R_{AC} = \frac{N_A}{N_C} = 1 - R_0. \therefore R_{AC} = 1 - 1.05 = -\frac{1}{20}$$

  Case B/C (A Fixed): From (8.14)

$$R_{BC} = \frac{R_0 - 1}{R_0}. \therefore R_{BC} = \frac{1.05 - 1}{1.05} = +\frac{1}{21}$$

- *Basic Ratio is* -1.05
  Case A/C (B Fixed): From (8.13)

$$R_{AC} = \frac{N_A}{N_C} = 1 - R_0. \therefore R_{AC} = 1 + 1.05 = +\frac{41}{20}$$

  Case B/C (A Fixed): From (8.14)

$$R_{BC} = \frac{R_0 - 1}{R_0}. \therefore R_{BC} = \frac{-1.05 - 1}{-1.05} = +\frac{41}{21}$$

8.8 Differential Trains

Common forms of differential trains are shown in Figures 8.10a and b. The general expression for speed ratios can be derived as follows:

1. First let member A be fixed. Then $N_C = N_B/R_{BC}$.
2. Next let member B be fixed. Then $N_C = N_A/R_{AC}$.
3. Now let A and B be free and rotate simultaneously, the total revolutions of member C will then be $N_C = N_B/R_{BC} + N_A/R_{AC}$.
4. Substitute these results of equations (8.13) and (8.14) in the above equation of step 3, which gives

$$N_C = \frac{N_B R_0}{R_0 - 1} + \frac{N_A}{1 - R_0} \tag{8.15}$$

5. Multiplying throughout by $(1 - R_0)$ and rearranging

$$N_A - N_B R_0 + N_C (R_0 - 1) = 0 \tag{8.16}$$

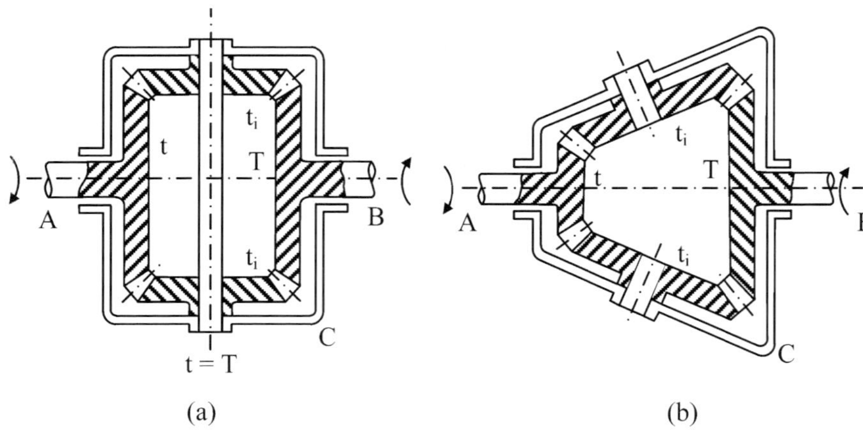

Fig. 8.10 Common forms of differential gear train

The differential train is a two-degree of freedom system, since you need two inputs to get a unique output or two output shafts are necessary for an input shaft.

Figure 8.10a is a common form of differential used in automobile transmission. Shafts A and B are coupled to the rear wheels and the planet carrier is driven by the propeller shaft (connected through a simple bevel gear). Since $T = t$, the basic ratio is -1 and equation (8.16) for this case becomes $N_A + N_B - 2N_C = 0$. Under normal conditions $N_A = N_B = N_C$. If the speed of one of the wheels increases due to slipping, the speed of the other wheel reduces, such that the sum of the two speeds of the wheels is equal to twice that of the drive shaft speed.

8.9 Torque Distribution in Epicyclic Trains

Besides the input and output torques, we have a holding torque on the fixed member in an epicyclic train. One condition is given from the equilibrium of the system

$$M_A + M_B + M_C = 0 \tag{8.17}$$

where M_A, M_B and M_C represent the torques on members A, B and C respectively.

The second condition is taken from the fact that the sum of the work done by all the three members is zero. Therefore

$$M_A N_A + M_B N_B + M_C N_C = 0 \tag{8.18}$$

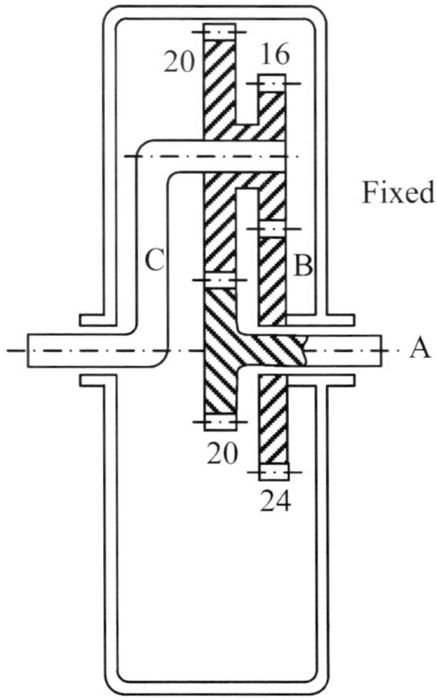

Fig. 8.11 Epicyclic train example

8.10 Example of an Epicyclic Train

Figure 8.11 shows an epicyclic train which consists of a planet carrier C driving the sun wheel A. The sun wheel B is fixed and the number of teeth on the gears is shown in the figure. Determine the speed of output shaft A, if C runs at 1500 RPM. Also determine the torques on all the members if the power transmitted is 10 HP.

1. First identify the three members of the basic epicyclic train as in Figure 8.9d. Figure 8.11 is the same as the case in Figure 8.8a, the planet carrier C here is made coaxial and the sun wheel shaft A runs inside the fixed shaft B. Remember that all epicyclic trains have the same elements, only their constructional features make them look different from one another. Once you make correct identification, solution of the problem becomes simple.
2. The basic ratio from equation (8.13a) is

$$R_0 = \frac{T_1 T_2}{t_1 t_2} = \frac{24 \times 20}{16 \times 20} = 1.5$$

3. The speed ratio is given by equation (8.14a), $R_{AC} = 1 - R_0 = -0.5$.
4. Therefore the speed of shaft A is $1500 \times (-0.5) = -750$ RPM.
5. The input torque is obtained from

$$\frac{2\pi \times 1500 \times M_C}{60 \times 746} = 10 \text{ HP}$$

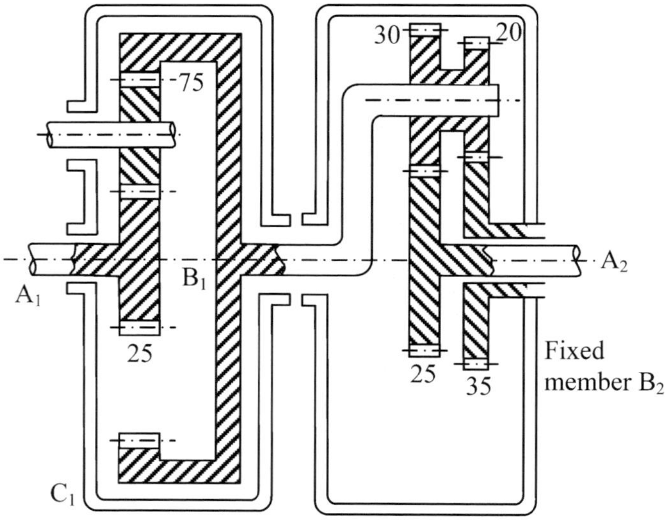

Fig. 8.12 Coupled epicyclic train

6. Or $M_C = 47.49$ Nm.
7. From equation (8.17), $M_A + M_B + 47.49 = 0$.
8. From equation (8.18), $M_A N_A + M_B N_B + 47.49 N_C = 0$.
9. Substituting in the above, we have $M_A(-750) + 47.49 \times 1500 = 0$ or $M_A = 94.98$ Nm.
10. Hence, $94.98 + M_B + 47.49 = 0$, i.e., $M_B = -142.47$ Nm.

8.11 Coupled Epicyclic Trains

In these trains, two or more of the simple epicyclic elements A, B or C appear in the coupled epicyclic trains. Ordinarily in a coupled epicyclic train made up of, say, two single trains, the two sets of elements A_1, B_1 and C_1 of the first train and A_2, B_2 and C_2 of the second train, must be present. An example of such a train is shown in Figure 8.12.

Determine the output speed of shaft A_2 for the input speeds of A_1 and C_1 given by -400 and $+200$ RPM respectively.

1. For the first train $R_{01} = -75/25 = -3$.
2. Equation (8.16) for the first train is

$$N_{A1} - N_{B1} R_{01} + N_{C1}(R_{01} - 1) = 0$$

$$-400 + 3N_{B1} + 200(-4) = 0$$

$$N_{B1} = 400 \text{ RPM}$$

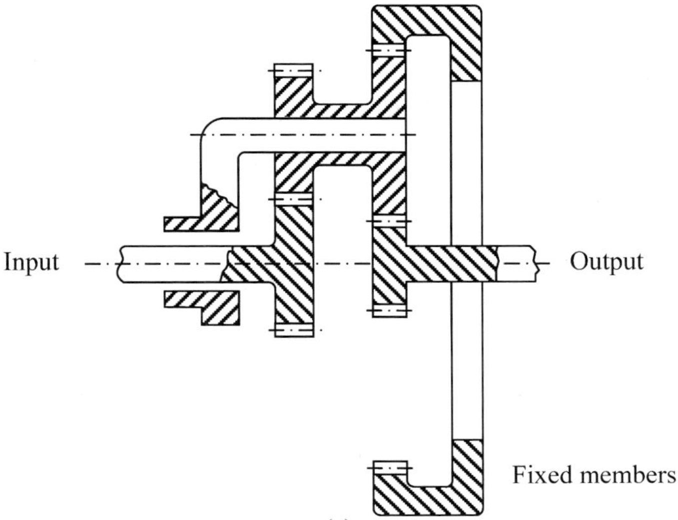

(a)

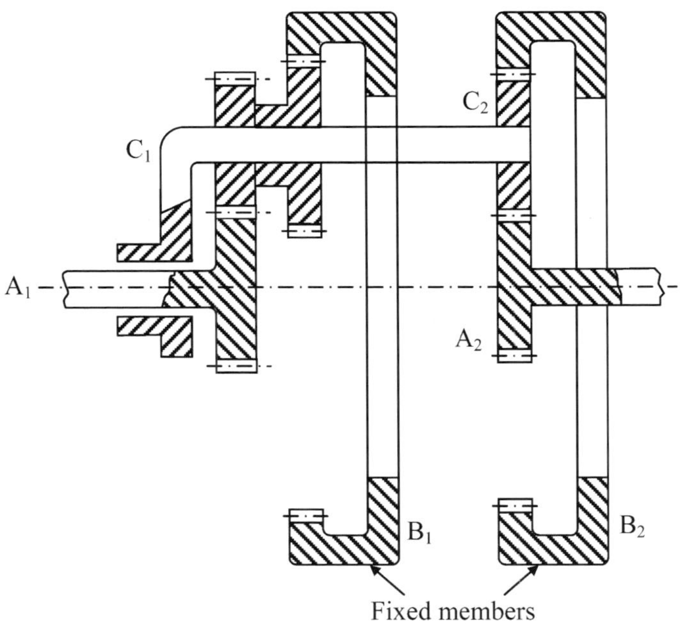

(b)

Fig. 8.13 Epicyclic train with three annular members and equivalent train

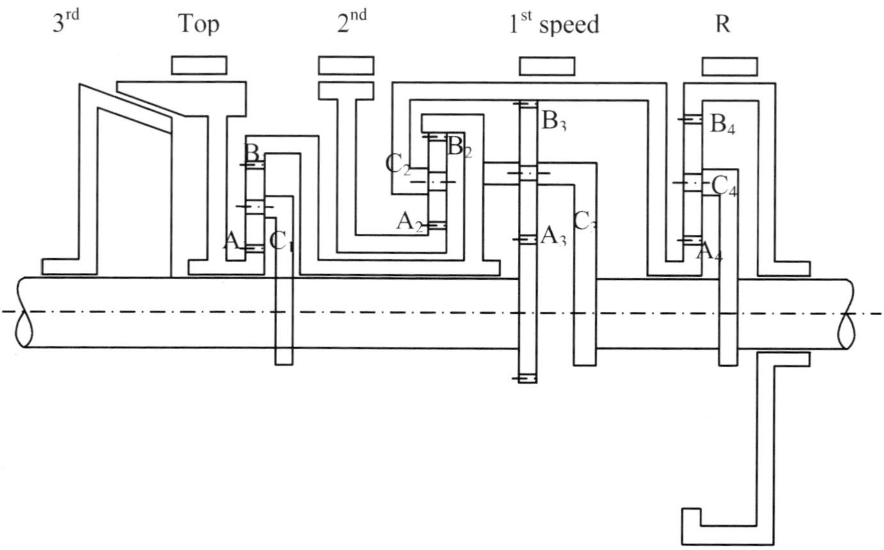

Fig. 8.14 Wilson four-speed automobile gear box

3. For the second train, $N_{C2} = N_{B1} = 400$ RPM and $N_{B2} = 0$.
4. The basic ratio for the second train is $R_{02} = \frac{30\times35}{20\times25} = 2.1$.
5. Applying (8.16) for the second train, $N_{A2} - N_{B2}R_{02} + N_{C2}(R_{02} - 1) = 0$, we get the speed of the output shaft $N_{A2} = -400 \times (2.1 - 1) = -400$ RPM.

Quite often, it may be difficult to identify the three sets of elements A_1, B_1 and C_1 of the first train and A_2, B_2 and C_2 of the second train. This is because; it is possible to design systems wherein, the job of one element can be duplicated by another one. In such cases, we have to imagine the presence of a member for the purpose of kinematic analysis. The following example illustrates this.

A compound epicyclic train is shown in Figure 8.13a. There are three annular members, the input shaft, output shaft and the fixed member in addition to the planet carrier. Find the two simple epicyclic trains that make up this compound train.

We can first identify the input shaft as A_1 and the planet carrier as C_1. If we ignore the output shaft in Figure 8.13a, the fixed annulus can be taken as fixed member B_1 of the first train. This leaves us apparently just one additional output shaft without two other members of the second train. Draw the first train separately, see Figure 8.13b and try to find a way of determining the second train members.

Imagine now A_2 also to be present, but separated from the first train. To have the second train, we need another planet carrier and a wheel or annulus. Extend the planet carrier C_1 to act as planet carrier for the second train also and choose the planet as a duplicate of the second gear in the gear block mating with the fixed annulus. All we need then is another identical fixed annulus as shown in Figure 8.13b.

In the train of Figure 8.13a, the annulus B_1 duplicates the work of fixed annulus B_2, so also the planet carrier C_1 functioning as planet carrier C_2. Therefore, to find the speed ratios of the train in Figure 8.13a, we first find all the elements that would have to be present in a compound train and perform the analysis.

Instead of duplicating the annulus B_1, an equivalent train, wherein the sun wheel A_1 is duplicated by another wheel A_2 to drive the second epicyclic train, can as well be chosen for the kinematic analysis of the problem in Figure 8.13a.

8.12 Wilson Four-Speed Automobile Gear Box

This gear box contains four coupled single epicyclic trains as shown in Figure 8.14. The individual elements of all the four elements are marked. Breaking pads on the elements B_4, B_3, A_2, A_1 and the clutch are used to give different speeds as explained below.

Reverse Speed. Figure 8.15a shows the effective train when B_4 is fixed. Here, trains 3 and 4 act together as a compound epicyclic train.

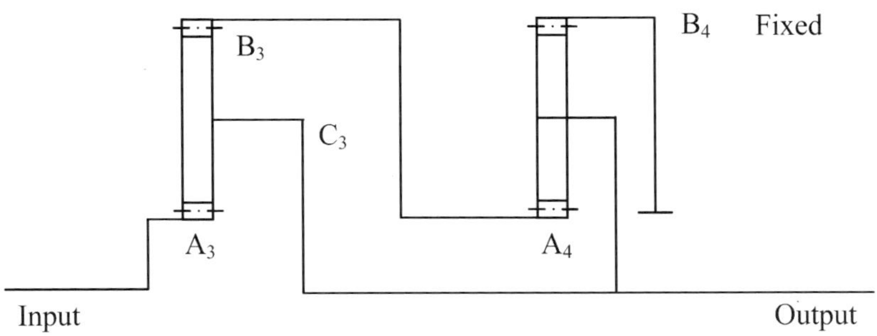

Fig. 8.15 (a) Reverse speed

First Speed. When B_3 is fixed, the train number 3 alone comes into operation as shown in Figure 8.15b.

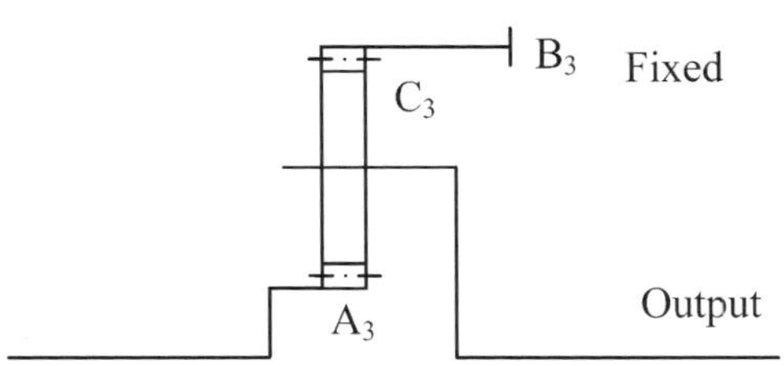

Fig. 8.15 (b) First speed

Second Speed. When A_2 is fixed, trains 2 and 3 come into operation as shown in Figure 8.15c.

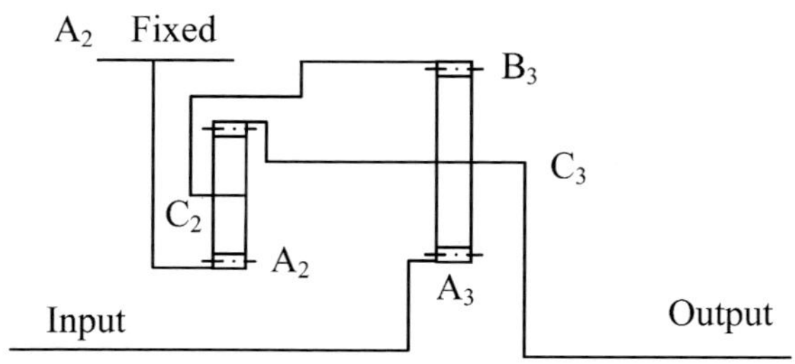

Fig. 8.15 (c) Second speed

Top Speed. When A_1 is fixed train 1 is only in operation as shown in Figure 8.15d.

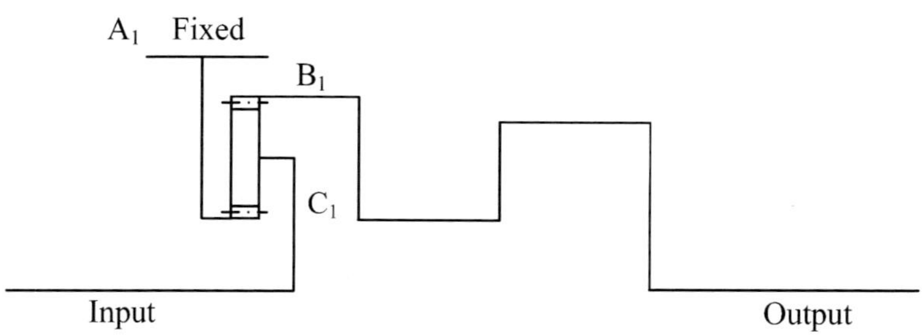

Fig. 8.15 (d) Top speed

Third Speed is obtained by pressing the clutch. This makes the sun wheel A_1 rotate at the speed as the input arm C_1. Further this necessitates that the planet carrier C_3 and the sun wheel A_3 rotate at the same speed. Hence all the trains 1 to 4 run en-bloc at the same speed and the output shaft will have the same speed as the input shaft.

8.13 Solved Problems

Solved Problem 8.1

A spur gear pair of DP 2 is to be used to connect two parallel shafts. The distance between the shaft axes is approximately 21.5 cm and the velocity ratio 4. Design the gear pair.

1. From the definition of diametral pitch, we have $t_1 = 2 \times d_1$ and $T_1 = 2 \times D_1$, where d_1, D_1 are the pitch diameters of the pinion and gear respectively.
2. From the given value of center distance, $(d_1 + D_1)/2 = 21.5$ cm.

3. Therefore, $t_1 + T = 2 \times 2 \times 21.5 = 86$.
4. We further have that $T_1 = 4t_1$.
5. Hence, $5t_1 = 86$. We choose the nearest number $t_1 = 17$.
6. Then, $T_1 = 68$.
7. The center distance actually obtained is $(17 + 68)/4 = 21.25$ cm.
8. You can choose the center distance to be 21.5 cm; however, the speed ratio will then not be 4. For example, if you choose $t_1 + T_1 = 86$ as in step 3, i.e., $T_1 = 69$, then $R = 4.06$.

Solved Problem 8.2

Design a compound gear train to give a total speed reduction of 250 to 1, in four steps. Use a DP ranging from 0.5 to 4. The center distances of the shafts should not incur decimal places. Make sure no interference takes place.

1. We first factorize the speed ratio,

$$250 = 4 \times 4 \times 15.625 = \frac{80}{20} \times \frac{80}{20} \times \frac{75}{20} \times \frac{75}{18}$$

2. For the first two stages with the same reduction, choose the diametral pitch to be 2.5.
3. Then the center distance in these two stages is

$$C_1 = C_2 = \frac{80 + 20}{2 \times 2.5} = 20 \text{ cm}$$

4. For the third stage, let the diametral pitch be 1.9, then

$$C_3 = \frac{75 + 20}{2 \times 1.9} = 25 \text{ cm}$$

5. Similarly for the fourth stage of reduction let the diametral pitch be 1.5 to give a whole number for the center distance,

$$C_4 = \frac{75 + 18}{2 \times 1.5} = 31 \text{ cm}$$

Solved Problem 8.3

Design a suitable gear train for the following speed ratios: (1) 5.3, (2) 7.9, (3) 2.09, (4) 5.39, (5) 11.83, (6) 13.409.

$$(1)\ \frac{53}{25}\times\frac{50}{20};\quad (2)\ \frac{79}{35}\times\frac{70}{20};\quad (3)\ \frac{33}{30}\times\frac{38}{20};$$

$$(4)\ \frac{49}{25}\times\frac{55}{20};\quad (5)\ \frac{39}{25}\times\frac{39}{18}\times\frac{63}{18};\quad (6)\ \frac{53}{20}\times\frac{55}{25}\times\frac{46}{20}$$

Solved Problem 8.4

Design a four-speed gear box with three indirect speeds for an automobile, to give speed ratios approximately 1.5 to 1, 2.5 to 1 and 4 to 1.

1. Let us assume a center distance of 15 cm and a module of 0.5 cm.
2. Let A be the gear mounted on drive shaft 1.
3. Let the gears B, C, E and G be all mounted in that order from the left on shaft 2.
4. Gears A and B are always in mesh.
5. The output shaft 3 carries three gears D, F and H in that order from the left side.
6. Shafts 1 and 2 can be coupled by a simple clutch to get the direct drive; in this position none other than gears A and B will be in mesh.
7. When the clutch is disengaged either gears D and C or F and E or H and G can be brought into mesh giving rise to third, second and first gears respectively.
8. I gear: $R = 4$.
9. $T_A + T_B = T_G + T_H = 2 \times 15 \times 2 = 60$.
10. Choose $T_A = 19$, $T_B = 41$, $T_G = 21$, $T_H = 39$.
11. The above teeth give a speed ratio of 4.0075. Note that a common factor between the teeth in each pair is avoided.
12. II gear: $R = 2.5$.
13. $T_E = 28$, $T_F = 32$ and $R = 2.47$.
14. III gear: $R = 1.5$.
15. $T_C = 35$, $T_D = 25$ and $R = 1.54$.

Solved Problem 8.5

A reverted train is to attain a speed ratio of 3.5. Design a train, using no wheel less than 15 teeth. The DP of the gears can be used in the range of 1 to 2.

Refer to Figure 8.4 of the text.

$$T_A = 50, \quad T_B = 20, \quad T_C = 35, \quad T_D = 25.$$

Solved Problem 8.6

In a brick making machine, a wide roller 25 cm dia mounted on shaft E is driven by a motor carrying a pulley A, which is 15 cm dia. The pulley in turn drives another pulley of 120 cm dia mounted on shaft B. On the same shaft B, a spur gear of 20 teeth is mounted, which meshes with a spur gear of 160 teeth mounted on shaft C.

The shaft C also carries another spur gear of 20 teeth which drives a 30 tooth gear on shaft E, through an idler of 112 teeth mounted on shaft D. Sketch the arrangement and determine the speed of conveyor belt running on the wide roller. The driving motor speed is 1450 RPM.

1. Overall speed ratio of the pulley and gear drive is

$$R = \frac{120}{15} \times \frac{160}{20} \times \frac{30}{20} = 96$$

2. The idler has no function in the speed ratio.
3. The speed of shaft E is

$$\frac{1450}{96} \times \frac{2\pi}{60} = 1.5817 \text{ rad/s}$$

4. Therefore, the speed of the conveyor belt is $1.5817 \times 12.5 = 19.77$ cm/s.

Solved Problem 8.7

In a simple lathe, the right-handed single threaded lead screw has 8 mm pitch. The pick-off gears available are 20, 25, 30, ..., 120 tooth. Sketch a tumbler train and determine a suitable gear train connecting the spindle and lead screw when (a) a right-hand screw thread with 10 threads per centimeter, and (b) a left-hand screw thread with 16 threads per centimeter have to be cut.

1. Refer to Figure 8.3. We consider gears A, D, E and F for the drive of the lead screw shaft carrying gear F.
2. Case (a): Pitch of lead screw is 0.8 cm.
3. Ten threads per cm (0.1 cm pitch) requires a speed ratio $R = 8$.
4. Therefore, choose from the pick-off gears, $T_A = 20$, $T_D = 40$, $T_E = 25$ and $T_F = 100$.
5. Case (b): Pitch of lead screw is 0.8 cm.
6. Sixteen threads per cm (0.0625 cm pitch) requires a speed ratio $R = 12.8$.
7. Since we require a left-handed thread, the idler is used to reverse the speed. Therefore, choose from the pick-off gears, $T_A = 25$, $T_D = 80$, $T_E = 20$ and $T_F = 80$.

Solved Problem 8.8

Design the gear box of a turret lathe to have a reduction of six speeds from 25 to 800 RPM. The main drive is a 1450 RPM AC motor. It is recommended that each stage ratio be less than 4.

1. $\phi = (800/25)^{1/(6-1)} = 2$.
2. Output speeds are 25, 50, 100, 200, 400 and 800 RPM.

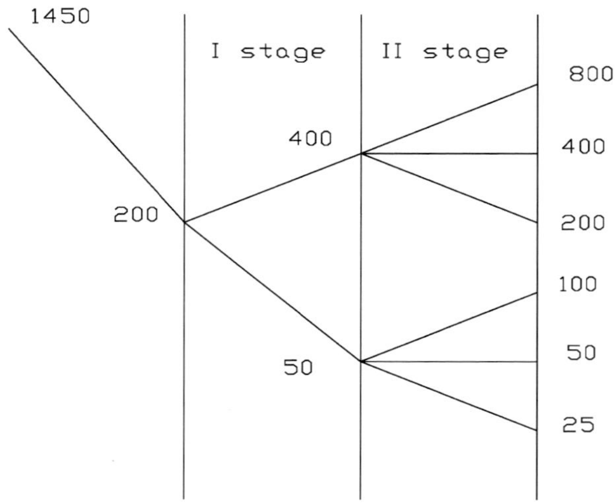

(a)

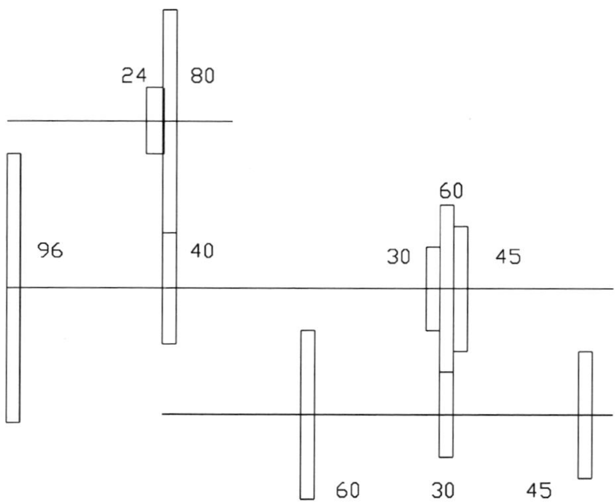

(b)

Fig. 8.16

3. You may adopt the kinematic layout shown in Figure 8.16a. Draw four vertical lines representing the input and output shafts and the two stages. Locate all six speeds at equal intervals.
4. The second stage consists of three speed ratios. From 400 RPM, draw lines representing the speed ratios of stage II to 800, 400 and 200 RPM. Draw parallel lines to these reductions from 50 RPM location to give 100, 50 and 25 RPM speeds on the output shaft.
5. Evidently stage I needs only two reduction lines. Let these reductions be from 200 RPM of the input shaft of stage I.
6. Use a belt drive to reduce the speed of the motor from 1450 RPM to 200 RPM.

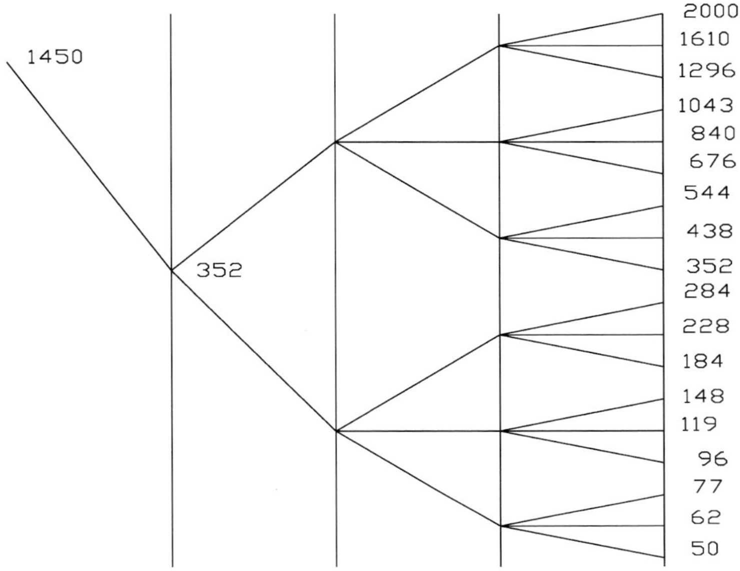

Fig. 8.17

7. Figure 8.16b shows the arrangement of gears.
8. Stage I: $T_A/T_B = 1/4$ and $T_C/T_D = 1/2$. Hence we may choose $T_A = 24$, $T_B = 96$, $T_C = 80$ and $T_D = 40$.
9. Repeat the above for stage II.
10. $T_E/T_F = 1/2$, $T_G/T_H = 1$ and $T_I/T_J = 2$. Hence $T_E = 30$, $T_F = 60$, $T_G = 45$, $T_H = 45$, $T_I = 60$ and $T_J = 30$.

Solved Problem 8.9

The head stock of a lathe is to be designed to give 18 speeds in the range of 50 to 2000 RPM. The main drive is a 1450 RPM AC motor. Sketch the head stock unit and the kinematic layout and determine the tooth numbers of all gears.

1. $\phi = (2000/50)^{1/(18-1)} = 1.2423353$.
2. This gives the output shaft speeds to be 50, 62 ($1.2423353 \times 50 = 62.117 \approx 62$), 77, 96, 119, 148, 184, 228, 284, 352, 438, 544, 676, 840, 1043, 1296, 1610 and 2000 RPM.
3. The ray diagram is shown in Figure 8.17. (Draw the headstock layout.) Three stages of reduction are used. The first stage has two speed ratios, stage II has three and the third stage another three making the 18 speeds possible as shown.
4. Stage I: $T_A/T_B = 1/2.96$ and $T_C/T_D = 2.386$.
5. Hence choose $T_A = 24$, $T_B = 71$, $T_C = 67$ and $T_D = 28$.

6. Actual speeds rounded to the first digit obtainable are 119 and 842 RPM respectively.
7. Stage II: $T_E/T_F = 1/1.92$, $T_G/T_H = 1$ and $T_I/T_J = 1.916$.
8. Hence $T_E = 26$, $T_F = 50$, $T_G = 38$, $T_H = 38$, $T_I = 50$ and $T_J = 26$.
9. Actual speeds rounded to the first digit obtainable are 62, 119, 228, 437, 840 and 1615 RPM.
10. Stage III: $T_K/T_L = 1/1.24$, $T_M/T_N = 1$ and $T_O/T_P = 1.242$.
11. Hence $T_K = 25$, $T_L = 31$, $T_M = 28$, $T_N = 28$, $T_O = 31$ and $T_P = 25$.
12. Actual speeds rounded to the first digit obtainable are 50, 62, 77, 96, 119, 148, 184, 228, 283, 353, 437, 542, 677, 840, 1041, 1302, 1615 and 2003 RPM.

Solved Problem 8.10

In an epicyclic train of the type shown in Figure 8.8b, the tooth numbers are $t_1 = 52$, $T_1 = 15$, $t_2 = 25$ and $T_2 = 92$. The planet carrier C rotates at 200 RPM and the input torque is 100 Nm. Determine the output speed of member A, if the internal wheel B is fixed. Also determine the output and holding torques.

1. From equation (8.13b), the basic ratio

$$R_0 = -\frac{T_1 T_2}{t_1 t_2} = -\frac{15 \times 92}{52 \times 25} = -1.06154$$

2. From equation (8.14a), the speed ratio of the train $R_{AC} = 1 - R_0 = 2.06154$.
3. Therefore the output shaft speed is $N_A = 200 \times 2.06154 = 412.3$ RPM.
4. From (8.18) $M_A N_A + M_B N_B + M_C N_C = 0$ gives $412.3 \times M_A + 200 \times 100 = 0$ or $M_A = -48.5$ Nm.
5. The holding torque is $M_B = -M_A - M_C = 48.5 - 100 = -51.5$ Nm.

Solved Problem 8.11

In a compound epicyclic train, the input shaft A_1 of the first train is connected to a sun wheel S_1 having 24 teeth. A planet wheel P_1 carried by the planet carrier C_1 drives an internal wheel B_1 having 66 teeth. The internal wheel B_1 is compounded with a sun wheel S_2 which has 28 teeth, while the planet carrier C_1 drives a pinion P_2 actuated on the sun wheel S_2 connecting a fixed internal wheel B_2 of 62 teeth. The planet carrier C_1 or C_2 forms the output shaft. Sketch the arrangement. If the input shaft speed is 1500 RPM and its torque is 500 Nm, determine the output shaft speed and torque and the holding torque.

1. Make a sketch first and identify the two epicyclic trains.
2. For the second train, the basic ratio is $R_{02} = -T_2/t_2 = -62/28 = -2.2143$.
3. Since B_2 is fixed, equation (8.14a) is applicable. $N_{A2}/N_{C2} = 1 - R_{02} = 3.2143$.
4. Therefore, $N_{A2} = 3.2143 N_{C2}$.

5. Since, as per the construction, $N_{A2} = N_{B1}$ and $N_{C2} = N_{C1}$, we have $N_{B1} = 3.2143N_{C1}$.
6. For the first train, the basic ratio from (8.13c) $R_{01} = -T_1/t_2 = -66/24 = -2.75$.
7. The first train is a differential as all three members are free to rotate. Therefore, using equation (8.16) $N_A - N_B R_0 + N_C(R_0 - 1) = 0$ we have $1500 + 2.75N_{B1} - 3.75N_{C1} = 0$.
8. Substituting $N_{B1} = 3.2143N_{C1}$, we get $1500 + 2.75 \times 3.2143N_{C1} - 3.75N_{C1} = 0$.
9. Therefore

$$N_{C1} = -\frac{1500}{2.75 \times 3.2143 - 3.75} = -294.8 \text{ RPM}$$

10. The output torque is

$$T_C = \frac{500 \times 1500}{-294.8} = -2545 \text{ Nm}$$

11. The holding torque is $T_{B2} = -2545 - 500 = -3045$ Nm.

Solved Problem 8.12

In the compound epicyclic train shown in Figure 8.13, the input wheel has 30 teeth and the output wheel 40 teeth. The fixed member has 80 teeth. The compound wheel has a 35 tooth gear meshing with the input wheel and a 20 tooth gear meshing with the output. Determine the speed ratio of the train.

1. Make the equivalent train as shown in Figure 8.13b.
2. Basic ratio of the first train

$$R_{01} = -\frac{T_1 T_2}{t_1 t_2} = -\frac{35 \times 80}{30 \times 20} = -4.6667$$

3. Basic ratio of the second train

$$R_{02} = -\frac{T_1}{t_1} = -\frac{80}{40} = -2.0$$

4. From the first train

$$\frac{N_{A1}}{N_{C1}} = 1 - R_{01} = 5.6667 = \frac{N_{A1}}{N_{C2}}$$

5. From the second train,

$$\frac{N_{A2}}{N_{C2}} = 1 - R_{02} = 3.0$$

6. Therefore, the speed ratio

$$\frac{N_{A2}}{N_{A1}} = \frac{3}{5.6667} = 0.529$$

Solved Problem 8.13

In the Wilson epicyclic gear box of the type shown in Figure 8.14, the following data are given:

$$T_{A1} = 30, \quad T_{B1} = 82$$
$$T_{A2} = 32, \quad T_{B2} = 68$$
$$T_{A3} = 38, \quad T_{B3} = 76$$
$$T_{A4} = 37, \quad T_{B4} = 77$$

Determine the four speed ratios:

1. *Reverse Gear: Trains 3 and 4 coupled together.*
   - $R_{03} = -76/38 = -2.0$.
   - $N_{A3} - N_{B3}R_{03} + N_{C3}(R_{03} - 1) = 0$, we have $N_{A3} + 2N_{B3} - 3N_{C3} = 0$.
   - $R_{04} = -77/37 = -2.081$.
   - $N_{B3} = N_{A4}$ and $N_{C3} = N_{C4}$.
   - $N_{A4}/N_{C4} = 1 - R_{04} = 1 + 2.081 = 3.08$.
   - $N_{A3} - N_{B3}R_{03} + N_{C3}(R_{03} - 1) = 0$ gives

$$N_{A3} + 2N_{A4} - \frac{3}{3.081}N_{A4} = 0$$

   - Therefore, $N_{A3} - 1.02N_{A4} = -3.143N_{C3}$.
   - The speed ratio for the reverse gear is $N_{C3}/N_{A3} = -0.318$.

2. *I Gear: Train 3.*
   - $R_{03} = -2.0$.
   - $N_{B3} = 0$.
   - $N_{A3} + N_{C3}(R_{03} - 1) = 0$ gives $N_{C3}/N_{A3} = 0.333$.

3. *II Gear: Trains 2 and 3 connected together.*
   - $R_{02} = -68/32 = -2.125$.
   - $N_{B2}/N_{C2} = 1.4706 = N_{C3}/N_{B3}$.
   - $N_{A3} + 2N_{B3} - 3N_{C3} = 0$, i.e., $N_{A3} + 1.363N_{C3} - 3N_{C3} = 0$.
   - $N_{C3}/N_{A3} = 0.61$.

4. *Top Gear: Gear train 1.*
   - $R_{01} = -82/30 = -2.7333$.
   - $N_{A1} - N_{B1}R_{01} + N_{C1}(R_{01} - 1) = 0$, i.e., $2.7333N_{B1} - 3.7333N_{C1} = 0$.
   - $N_{B1}/N_{C1} = 1.366$.

8.14 Additional Problems

1. Design a compound gear train for an exact train ratio of 180:1.
2. Design a reverted compound train for a speed ratio 18:1.
3. In a simple epicyclic train, the sun wheel has 40 teeth, the planet has 20 teeth and the annulus has 80 teeth. The planet carrier is driven at 200 RPM in a clockwise direction and the sun wheel gets an independent drive at 100 RPM also in a clockwise direction. What is the output shaft speed?
4. An epicyclic train is designed to have a common planetary wheel of 20 teeth operating on three sun wheels of 100, 99 and 101 teeth placed side by side in the order given. The center distances between all sun gears and the planet are kept the same (this is possible because of involute tooth properties). The planet carrier is given an input of 100 RPM in a counter-clockwise direction and the sun wheel of 100 teeth is fixed. Show that the other two wheels rotate slowly at just about 1 RPM in opposite directions.
5. A differential gear train consists of a planet carrier carrying two planets, a sun wheel and an annulus all rotating about a common axis. The sun wheel has 30 teeth and meshes with one of the two gears in the first planet block which has 45 teeth. The second gear of this planet block has 25 teeth and meshes with the second planet wheel (carried by the same planet carrier arm) having 50 teeth. The second planet wheel is a single wheel and meshes in turn with the annulus having 200 teeth. If the planet carrier arm rotates at 50 RPM in a clockwise direction and simultaneously the annulus rotates at 20 RPM in a counter-clockwise direction, what is the speed of the sun wheel?
6. In an epicyclic gear train of the type in Figure 8.13a, the input sun wheel has 30 teeth and the output sun wheel 40 teeth. The input sun wheel mates with the planet wheel of 35 teeth and the planet wheel mating with the output wheel has 20 teeth. The fixed annulus has 100 teeth. Determine the speed ratio of the train.

Index